Autorenkollektiv

# Synthetic and Structural Problems

Organotitanium Reagents in Organic Synthesis — A Simple Means to Adjust Reactivity and Selectivity of Carbanions
M. T. Reetz

Lithium Halocarbenoids — Carbanions of High Synthetic Versatility
H. Siegel

Pyridinophanes, Pyridinocrowns, and Pyridinocryptands
V. K. Majestic, G. R. Newkome

New Perspectives in Polymer-Supported Peptide Synthesis
V. N. R. Pillai, M. Mutter

AKADEMIE-VERLAG · BERLIN

Autorenkollektiv

Synthetic and Structural Problems

Autorenkollektiv

# SYNTHETIC AND STRUCTURAL PROBLEMS

With 3 Figures and 13 Tables

AKADEMIE-VERLAG · BERLIN
1983

Die Originalausgabe erscheint im Springer-Verlag Berlin · Heidelberg · New York

Vertrieb ausschließlich für die DDR und die sozialistischen Länder



Erschienen im Akademie-Verlag, DDR-1086 Berlin, Leipziger Straße 3–4
Lizenznummer: 202 · 100/552/83
Gesamtherstellung: VEB Druckerei „Thomas Müntzer“, 5820 Bad Langensalza
Umschlaggestaltung: Karl Salzbrunn
Bestellnummer: 763 222 4 (6758) · LSV 1275
Printed in GDR
DDR 86,— M

# Table of Contents

**Organotitanium Reagents in Organic Synthesis A Simple Means to Adjust Reactivity and Selectivity of Carbanions**
M. T. Reetz . . . . . . . . . . . . . . . . . . . . . . 1

**Lithium Halocarbenoids — Carbanions of High Synthetic Versatility**
H. J. Siegel. . . . . . . . . . . . . . . . . . . . . . 55

**Pyridinophanes, Pyridinocrowns, and Pyridinocryptands**
V. K. Majestic, G. R. Newkome . . . . . . . . . . . . 79

**New Perspectives in Polymer-Supported Peptide Synthesis**
V. N. R. Pillai, M. Mutter . . . . . . . . . . . . . . . . 119

# Organotitanium Reagents in Organic Synthesis
# A Simple Means to Adjust Reactivity and Selectivity of Carbanions

**Manfred T. Reetz**

Fachbereich Chemie der Universität Marburg, Hans-Meerwein-Straße, 3550 Marburg, FRG

## Table of Contents

A. **Introduction** . . . . 3

B. **Synthesis, Structure and Properties of Some Common Organotitanium Compounds** . . . . 5
B.I Synthesis, Stability and Aggregation State . . . . 5
B.II Bond Lengths and Angles . . . . 8
B.III Bond Energies and the Nature of the C—Ti σ-Bond . . . . 8

C. **Increasing Chemoselectivity of Carbanions by Conversion into Titanium Reagents** . . . . 9
C.I General Remarks . . . . 9
C.II Discrimination between Aldehydes and Ketones . . . . 10
C.III Types of Additional Functional Groups Tolerated . . . . 13
C.IV Discrimination between Compounds belonging to the same Functional Group . . . . 16
C.V Addition to Enolizable Ketones . . . . 18
C.VI. Relative Rates . . . . 19
C.VII Controlled Reversal of Chemoselectivity Using Titanium Ate Complexes 20

D. **Diastereoselectivity in Reactions of Organotitanium Reagents with Carbonyl Groups** . . . . 24
D.I Asymmetric Induction in Acyclic Systems . . . . 24
D.II Diastereoselectivity in Addition Reactions of Cyclic Ketones . . . . 31
D.III Diastereoselective Aldol-Type Additions . . . . 35
D.III.1 Titanium Enolates in Aldol Additions . . . . 35
D.III.2 Aldol-Type Additions of Titanated Hydrazones . . . . 38
D.III.3 Addition of Substituted Allyltitanium Reagents to Aldehydes and Ketones . . . . 40

| | | |
|---|---|---|
| E. | **Enantioselective Addition of Chirally Modified Organotitanium Reagents to Carbonyl Compounds** | 42 |
| F. | **Reaction Types** | 43 |
| F.I | Methylation of Tertiary Alkyl Halides | 43 |
| F.II | Direct Geminal Dimethylation of Ketones | 45 |
| F.III | Direct Geminal Dialkylation of Ketones | 45 |
| F.IV | Methylation of Secondary Alkyl Halides | 47 |
| F.V | Wittig-Type Olefination | 47 |
| G. | **Hints on how to Use Organotitanium Reagents** | 48 |
| H. | **Conclusions and Further Work** | 48 |
| I. | **References** | 50 |

**List of Abbreviations**

Me = methyl
Et = ethyl
Pr = n-propyl
$^{i}$Pr = isopropyl
Bu = n-butyl
Ph = phenyl
Cp = $\eta^5$-cyclopentadienyl
M = metal

Classical organometallic reagents such as alkyl-lithium and magnesium compounds as well as a large number of lithiated species which are substituted by resonance-stabilizing groups or heteroatoms can be converted into the titanium analogs using cheap quenching reagents. This increases chemoselectivity dramatically and also has a profound influence on stereoselectivity in reactions with carbonyl compounds. Since the nature of the ligands at titanium can ge varied systematically, it is possible to control the electronic and steric environment at the metal in a predictable manner. Besides increasing selectivity, several novel reaction types are possible.

# A. Introduction

Transition metal organometallic chemistry is turning out to be a powerful tool in synthetic organic chemistry. A large number of reagents containing such metals as Fe, Ni, Cu, Zr and Pd are being used for chemo-, regio- and stereoselective C—C bond formation. It is surprising that titanium has lagged somewhat behind in this rapidly expanding field, particularly in view of the fact that it is cheap and that a large number of organotitanium reagents have already been prepared [1]. Most of them were synthesized by research groups interested in Ziegler-Natta-type polymerization [2] or nitrogen fixation [3]. Furthermore, the participation of organotitanium or zirconium compounds in a number of interesting processes such as CO [4] and $SO_2$ [5] insertion, olefin metathesis [6], reduction [7], hydrometallation [8] and carbometallation [9] has been reported. In most of these and other reactions titanium bears two π-bonded cyclopentadienyl ligands (totanocene derivatives) [10]. In many of the reactions zirconium is better suited, as in much of the chemistry described by Schwartz [8] and Negishi [9]. Although not organometallic in nature, three other titanium compounds deserve mention. McMurry's reagent ($TiCl_3/LiAlH_4$) and related systems induce the reductive dimerization of aldehades and ketones [11]. Secondly, $TiCl_4$ is a mild Lewis acid which mediates aldol and Michael addition of silyl enol ethers as described by Mukaiyama [12]. O-silylated carbonyl compounds also react with $S_N1$ active alkyl halides in the presence of $TiCl_4$ [13], a process which is complementary to traditional alkylations of basic Li- and Mg-enolates in which only $S_N2$ active alkylating agents can be used. $TiCl_4$ is also an excellent Lewis acid in reactions of allyl, aryl and vinyl silanes with a variety of electrophiles [14]. Thirdly, the patent literature describes several uses of titanium tetraalkoxides, $Ti(OR)_4$, in organic reactions, e.g., as mild transesterification catalysts [15,16]. In another application, titanium tetraisopropoxide mediates the liquid-phase epoxidation of allyl alcohols [17].

A particularly exciting advancement in this field is the enantioselective version as reported by Sharpless [18]. Since several reviews [19a–f] and books [19g–j] concerning most of the above reactions as well as additional aspects of organotitanium chemistry have appeared, only brief reference will be made to them here.

*The purpose of the present review is to focus on a new development in organotitanium chemistry which promises to be useful in synthetic organic chemisty.* The type of C—C bond formation to be discussed is generally *not* based on typical transition metal behavior such as oxidative coupling, CO insertion, β-hydride elimination, etc. Rather, reaction types traditional to "carbanion" chemistry [20] are most often involved, e.g., Grignard-type addition, aldol consensation, Michael addition, and substitution reactions of alkyl halides. We propose that reactivity and selectivity of many classical "carbanions" and organometallic reagents such as RMgX, RLi, RZnX as well as resonance stabilized and hetero-atom substituted species can be modified in a synthetically meaningful way by conversion into the corresponding titanium analogs [21] (Scheme 1)[1].

---

[1] First reported in a lecture at the Max-Planck-Institut für Kohleforschung, Mülheim on Dec. 10, 1979

Scheme 1: Variable Adjustment of Carbanion Reactivity and Selectivity by Conversion into Titanium Reagents

$$R{-}Li \xrightarrow{ClTiX_3} R{-}TiX_3$$

R = Organic Moiety
X = Cl, OR′, $NR'_2$, etc.

The above transformation has two different goals [21,22]: 1) To increase chemo-, regio-, diastereo- and enantioselectivity in the reaction with carbonyl compounds (Sections C–E), and 2) to make certain reaction types amenable which do proceed readily with classical reagents, e.g., methylation of tertiary alkyl halides, alcohols and ethers, and direct geminal dialkylation of ketones (Section F). It turns out that organotitanium compounds are usually complementary to Li, Mg, Zn, Fe, Ni, Cu and Pd reagents. So far, experience in the above two areas points to the following *positive* aspects:

1) The electronic property at titanium, e.g., Lewis acidity, can be gradually varied in a predictable way by varying the nature of the ligand X (Scheme 1). Since this also affects Lewis basicity and nucleophilicity of the organic moiety, *variable adjustment of reactivity and selectivity is possible.*
2) The steric environment within a certain series can be changed by increasing the size of the R′ group in the ligand OR′ or $NR'_2$ (Scheme 1).
3) Chirally modified titanium reagents are accessible.
4) The σ-bonds to titanium are covalent, and bond distances and angles are known, making the interpretation of certain reactions more easy relative to those involving classical reagents such as RLi.
5) Many organotitanium reagents are monomeric, again simplifying mechanistic discussions.
6) The solubility of organotitanium reagents is generally much higher than that of the lithium or magnesium analogs, so that such solvents as pentane, benzene, dichloromethane, ether, THF, etc. can be chosen.
7) Many of the quenching reagents (Scheme 1) are easily accessible using cheap, commercially available materials. A prime example is the synthesis of chlorotitanium triisopropoxide *3* from titanium tetrachloride *1* and titanium tetraisopropoxide *2*. Although solvents can be used for the reaction [23,24], simply mixing the components in the right proportions at room temperature under an inert atmosphere for several hours results in quantitative conversion [25], distillation at 0.8 torr/86 °C not being necessary; the compound solidifies at room temperature. In our laboratories we have prepared kg batches by this method (Equation 1).

$$\underset{1}{TiCl_4} + 3\,\underset{2}{Ti(OCHMe_2)_4} \rightarrow 4\,\underset{3}{ClTi(OCHMe_2)_3} \qquad \text{(Eq. 1)}$$

8) Workup does not afford highly toxic materials as is known for reactions involving such metals as Cd, Sn or Hg.

*Negative* aspects are as follows:
1) Secondary and tertiary alkyltitanium compounds are generally not accessible due to decomposition via β-hydride elimination.
2) Certain vinyltitanium compounds are also unstable, in this case the undesirable transition metal-like behavior being oxidative coupling.
3) Carbanions which are particularly electron rich cannot be titanated smoothly with quenching reagents having chloro or alkoxy groups at titanium due to reduction of Ti(IV) to Ti(III) via single electron transfer (SET).
4) Some organotitanium reagents are *not* monomeric, thereby clouding mechanistic interpretations.
5) A number of organotitanium compounds are so tempered in their reactivity, that they react sluggishly or not at all.
6) Titanium compounds having hetero-atoms readily undergo redistribution or ligand exchange reactions, thereby making the synthesis of optically active derivatives having the center of chirality at *titanium* difficult.
7) In isolated cases workup is hampered by the formation of $TiO_2$-containing emulsions.

Some but not all of these problems have been solved (Sects. C–G). Before presenting the chemistry of organotitanium compounds, the synthesis, structure and properties of some simple derivatives will be considered because the data is useful in interpreting reactivity and selectivity.

# B. Synthesis, Structure and Properties of Some Common Organotitanium Compounds

## B.I Synthesis, Stability and Aggregation State

For a long time it was thought that binary complexes of transition metals containing only σ-alkyl- or aryl ligands could be thermodynamically stable only if they are coordinatively saturated (18 electron rule). A series of illuminating papers by Wilkinson [26], Lappert [27] and others [28] dispelled this fallacy. It became clear that transition metal alkyls can decompose via *kinetically* low energy pathways such as β-hydride elimination or abstraction, but rarely via *primary* metal-carbon homolysis. Mechanisms and theories of decomposition have been proposed and reviewed [29].

30 years ago Herman and Nelson isolated the first organometallic titanium compounds incorporating a Ti—C σ-bond by reacting titanium tetraisopropoxide *2* with phenyllithium (containing LiBr) and treating the structurally undefined adduct *4* with $TiCl_4$ to produce phenyltitanium triisopropoxide *5* (40% overall yield) [30a] (Equation 2). Later, considerably better procedures using *3* were developed [1]. The light yellow compound is indefinitely stable below 10 °C, but decomposes if heated above its melting point (88–98 °C) to form violet colored Ti(III) species. These authors also synthesized other aryl, ethinyl, methyl and n-butyl derivatives without

isolating them. The ease of decomposition was stated to increase in this sequence [30b)].

$$2 \xrightarrow[\text{LiBr}]{\text{PhLi}} PhTi(OCHMe_2)_4 \cdot Li_2Br \cdot O(C_2H_5)_2 \xrightarrow{\text{TiCl}_4} PhTi(OCHMe_2)_3 \quad \text{(Eq. 2)}$$

*4* *5*

It was not until the pioneering studies of Bestian, Clauss and Beermann at Farbwerke Hoechst, that a number of methyltitanium alkoxides and chlorides were prepared in high yields, isolated and characterized [31)]. Later, additional examples and improvements were reported [1, 19, 28, 32)]. Methyltitanium triisopropoxide *6* is readily prepared by quenching methyllithium with *3*, decanting from the inorganic precipitate and distilling at 0.01 torr/50 °C (95% yield) [31b, 32)].

$$3 \xrightarrow{\text{CH}_3\text{Li}} CH_3Ti(OCHMe_2)_3 \quad \text{(Eq. 3)}$$

*6*

Cryoscopic measurements of *6* in benzene are in line with a monomeric form [33)]. However, an osmometric study in the same solvent indicates a small degree of association [32)]. In contrast, methyltitanium triethoxide 7 is dimeric, as shown by cryoscopic studies [33)]; accordingly, even at 80 °C/0.01 torr it fails to distill [31b)]. Another example of a dimeric structure pertains to the dimethyl derivative *8* in solution [34a)] as well as in the solid state [34b)]. The X-ray structure determination shows dimerization to occur through oxygen bridges, pentavalent titanium being bonded to three oxygens and two carbons in a distorted trigonal-bipyramidal configuration [34b)]. These phenomena are closely related to the different degrees of aggregation of titanium tetraalkoxides, $Ti(OR)_4$, which have been studied in great detail as a function of the bulkiness of the alkoxide group [19f)]. Straight-chain n-alkoxides are trimeric in benzene, while sterically more hindered derivatives (e.g., isopropoxides) are monomeric [19f)]. We shall see that differences in the rate of reaction of methyltitanium trialkoxides with carbonyl compounds relate to different degrees of aggregation (Sect. C.V).

$CH_3Ti(OEt)_3$ — *7*

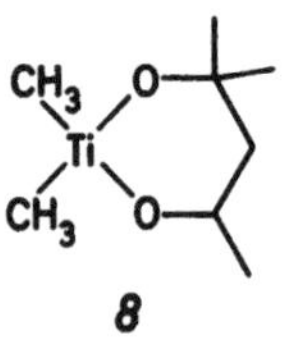


*8*

$CH_3CH_2Ti(OCHMe_2)_3$ — *9*

$CH_3CH_2CH_2CH_2Ti(OCHMe_2)_3$ — *10*

Homologs of *6*, e.g., ethyl and n-butyl derivatives *9* and *10* are more sensitive, but can be handled in solution [21, 30c)]. However, this does not apply to branched derivatives such as isopropyltitanium triisopropoxide, which appear to be unstable even in solution due to β-Hydride elimination. Moderate to pronounced degrees of thermal stability are characteristic of compounds in which β-hydrogens are not easily abstracted due to strain in the corresponding olefin (e.g., *11* [35)] or *12* [36)] or in which they are lacking (e.g., *13* [37)]). Tetramethyltitanium *14* has no β-hydrogen atoms but is nevertheless unstable above —20 °C, alternative modes of decomposition setting in [38)].

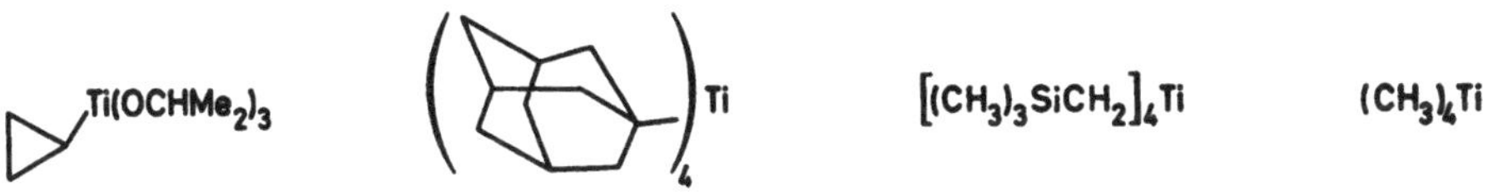


*11* *12* *13* *14*

Bürger et al. have shown that the presence of amino ligands at the metal enhances the thermal stability of primary, secondary and tertiary alkyl as well as allyl and vinyl titanium compounds dramatically [39)]. For example, n-propyltitanium tris(diethylamide) *16* can be isolated in pure form by sublimation at 60 °C/$10^{-3}$ torr (68 %) [39b)].

(4)

*15* *16*

Methyltitanium trichloride *17* is accessible in high yield by treating $TiCl_4$ with a variety of methylating agents such as $CH_3MgX$, $CH_3Li$, $CH_3AlCl_2$ or $(CH_3)_2Zn$ [1)]. The latter reagent allows for the high yield preparation of *17* in non-etheral solvents such as pentane [40)] or methylene chloride [41)]. *17* can be distilled or crystallized (m.p. 29 °C) and is monomeric in the gas phase [42)] and in solution [31b)]. In $CH_2Cl_2$ it is stable for weeks in the refrigerator; crystalls decompose. Homologs are rather labile [1,19)]. π-bonded cyclopentadienyltitanium compounds [1b)] are thermally very stable, e.g., monomeric *18* [43)].

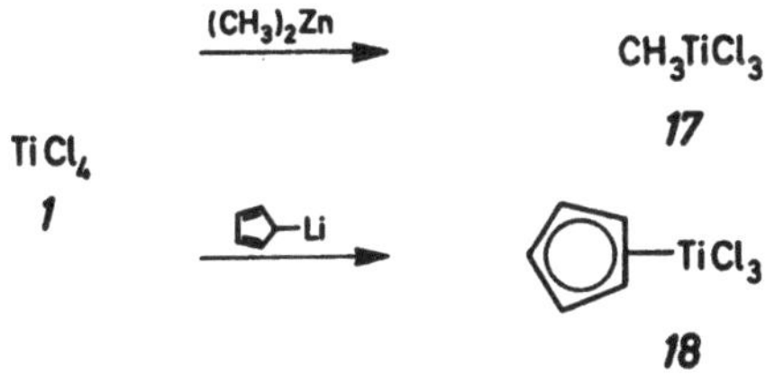


(5)

(6)

An important property of many organotitanium reagents is their ability to form 1:1 trigonal bipyramidal (five-coordinate) or more commonly 1:2 octahedral (six-coordinate) adducts with donor ligands such as ether, THF, dioxane, glyme, pyridine, diamines, 2.2′-bipyridyl, thioethers and phosphines [1,19)]. Generally, this increases kinetic stability and allows for the isolation of otherwise labile compounds, e.g., the 2,2′-bipyridyl adduct of tetramethyltitanium as described by Thiele [38c,44)]. An exception appears to be the ether-adduct of methyltitanium trichloride *17*, since decomposition is *faster* than *17* itself [1b,45)].

## B.II Bond Length and Angles

Monomeric titanium(IV) compunds are tetrahedral, although distortions have been reported [1,19]. A case in point is tetrabenzyltitanium, in which the mean TiC—C bond angle is 103°, bringing the face of the phenyl group closer to the metal nucleus [46]. This has been ascribed to an interaction between the aromatic π-electrons and the d orbitals of the metal. Deviation from ideal tetrahedral geometry is even greater in tetrabenzylzirconium, but not in the analogous tin compound [47]. Related interactions have been proposed to account for unexpected diastereoselectivities in certain reactions (Sect. C). Table 1 summarizes some typical bond lengths [1,19,48] for Ti, Zr and other metals. The data is of considerable use in discussing certain reactions (Sect. D.I).

**Table 1.** Typical Bond Lengths [1,19,48]

| Metal | Metal-Carbon Bond Length (Å) | Metal-Oxygen Bond Length (Å) |
|---|---|---|
| Ti | ~2.10 | 1.68–1.78 |
| Zr | ~2.20 | 2.10–2.15 |
| Li | ~2.00 | 1.92–2.00 |
| Mg | ~2.00 | 2.00–2.13 |
| B | 1.51–1.58 | 1.36–1.47 |

## B.III Bond Energies and the Nature of the C—Ti σ-Bond

Thermodynamic data in the area of transition metal chemistry is available, but additional studies would be desirable. One of the early indications that C—Ti bonds are *not* notoriously weak was obtained from the heats of combustion of $Cp_2Ti(CH_3)_2$ and $Cp_2Ti(C_6H_5)_2$ with subsequent estimation of the σ-bond dissociation energies (250 $kJ/mol^{-1}$ and 350 $kJ/mol^{-1}$, respectively) [49]. From heats of alcoholysis of a number of titanium, zirconium and hafnium compounds, and heats of solution of the products as well as subsidiary data, Lappert estimated heats of formation ($\Delta H_f^\circ$) and thermochemical mean bond energy terms ($\bar{E}_{M-X}$) of metal—X bonds [50] (Table 2).

**Table 2.** Thermochemical Data [50] of some Complexes $MX_4$

| Compound | Bond (M-X) | $\bar{E}$ (M-X) kcal/mol |
|---|---|---|
| $TiCl_4$ | Ti—Cl | 102 |
| $Ti(CH_2SiMe_3)_4$ | Ti—C | 64 |
| $Ti(CH_2CMe_3)_4$ | Ti—C | 44 |
| $Ti(CH_2Ph)_4$ | Ti—C | 63 |
| $Ti(NMe_2)_4$ | Ti—N | 81 |
| $Ti(OCHMe_2)_4$ | Ti—O | 115 |
| $ZrCl_4$ | Zr—Cl | 117 |
| $Zr(CH_2SiMe_3)_4$ | Zr—C | 75 |
| $Zr(NMe_2)_4$ | Zr—N | 91 |

The results show that mean bond strengths for Ti, Zr and Hf decrease in the sequence M—O > M—Cl > M—N > M—C, and that the values are monotonically *higher* for Zr than Ti compounds by about 15%. Interestingly, for compounds with neopentyl groups a considerably weakened M—C bond is found, which has been attributed to steric conjestion [27, 50]. In any case, the metal-carbon bonds are *not* particularly weak relative to M—X bonds involving main group metals [51]. Also, a high thermodynamic driving force to form M—O bonds in organic reactions is to be expected.

The electronic configuration of titanium is [Ar] $3d^2 4s^2$, which means that Ti(IV) compounds are $d^0$ species with free coordination sites [1, 27, 28]. $^1$H-NMR and $^{13}$C-NMR data are known and have been occasionally discussed in terms of bond polarity [1, 19], but such interpretations are obviously of limited value. The electronic structure of methyltitanium trichloride *17* and other reagents have been considered qualitatively [52] and quantitatively [53–56] using molecular orbital procedures. It is problematical to compare these calculations in a quantitative way with those that have been carried out for methyllithium [57] since different methods, basis sets and assumptions are involved, but the extreme polar nature of the C—Li bond does not appear to apply to the C—Ti analog. Several MO calculations of the π-interaction between ethylene and methyltitanium trichloride *17* (models for Ziegler-Natta polymerization) clearly emphasize the role of vacant coordination sites at titanium [58].

Photoelectron spectra of a few organotitanium compounds provide additional insight into the bonding [59, 60]. For example, in $CH_3TiX_3$ (X = OR, $NR_2$) back-donation of lone electron pairs explains some of the spectral peculiarities [60].

π-Donation of heteroatoms was also considered in compounds of the type $Cp_2TiX_2$ (X = Cl, OR) by Caulton [61]. On the basis of crystal structural and thermodynamic data alkoxy groups were postulated to be effective π-donors. Additional evidence for $p_\pi$—$d_\pi$ interaction between titanium(IV) and heteroatoms is also available from spectroscopic data [1, 19] (e.g., NMR-data [62a]) and from studies concerning Lewis acid/Lewis base interactions [19f–g, 62b]. The fact that it is very difficult to separate electronic and geometric factors has been overlooked on occasion. Nevertheless, the π-donor ability seems to increase in the series Cl < OR < $NR_2$, which means that Lewis acidity decreases in this order [1, 19, 62]. As we shall see, this has important ramifications concerning reactivity and selectivity of organotitanium compounds.

## C. Increasing Chemoselectivity of Carbanions by Conversion into Titanium Reagents

### C.I General Remarks

Grignard and alkyllithium reagents as well as a host of resonance-stabilized and hetero-atom substituted carbanions add smoothly to carbonyl compounds, the reactions being of fundamental importance in synthetic organic chemistry [20, 63, 64]. Unfortunately, in systems containing several functional groups, chemoselectivity is frequently low. The simplest case refers to Grignard reagents, which often fail to discriminate effectively between aldehydes, ketones and other functional groups [64a].

For example, Kharasch found that phenylmagnesium bromide reacts with acetone slightly faster than with acetaldehyde ($k_{rel} \simeq 1.4$) [65a)]. In other cases, aldehydes are somewhat more reactive [65b)]. The polar and highly reactive alkyllithium compounds [64b)] are even less selective [66)] (Equation 7).

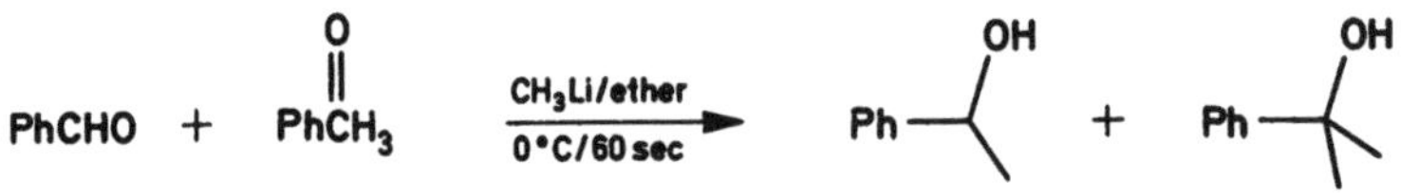


*19* *20* *21* *22*

1 : 1 (7)

Organozinc [67)] and cadmium [68)] compounds have been used to convert acid chlorides into ketones, but in most cases reactions with aldehydes or ketones are too sluggish to be of synthetic use. Not much is known concerning Grignard-type addition of cuprates, although there are indications that aldehydes react faster than ketones [69)]. A few alkylmanganese compounds have been reacted with acid chlorides in a synthetically important reaction, the products (ketones) remaining inert under the conditions [70a)]. Very recently, several methyl transition metal compounds have been added selectively to aldehydes [70b)].

## C.II Discrimination between Aldehydes and Ketones

Although several organotitanium compounds were shown to give a positive Gilman test [1,19)], synthetic implications did not become apparent until 1979/80 [71)]. We noticed that the addition of a variety or organotitanium reagents to aldehydes occured smoothly at low temperatures (—78 to —10 °C), 1—2 h, >90% conversion) [72)]. In contrast, the analogous reaction with ketones required room temperature (8–72 h, 80–90% conversion) or refluxing in THF. This as well as other observations led to the general working hypothesis that organotitanium compounds are highly chemoselective reagents [21,72–75)]. In a related study Seebach reached similar conclusions [76)]. Early studies revealed that phenyltitanium triisopropoxide *5* [72)], methyltitanium triisopropoxide *6* [72,76)] and methyltitanium trichloride *17* [72)] react with 1:1 mixtures of aldehydes and ketones to form essentially only the *aldehyde* adduct, e.g., Equation 8. In other examples, *6* (one part) or *17* (one part) reacts with a mixture of 3-methylbutanal (one part) and 2-heptanone (one part) to afford only the aldehyde adduct [72)]. *6* also adds selectively to benzaldehyde in the presence of acetophenone [76,77)], in contrast to methyllithium[2] (Eq. 7) or methylmagnesium-bromide [66,76)].

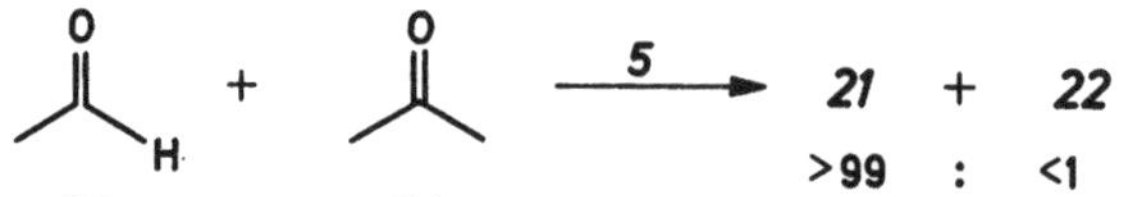


*23* *24* (8)

[2] A 1:1 mixture of benzaldehyde and acetophenone reacts with 1 part methyllithium in ether (0 °C, 60 sec) to form a 1:1 mixture of aldehyde and ketone adducts with little side products [66)]. Under slightly different reaction conditions a 2:3 ratio is obtained [76)].

In subsequent investigations the generality of the above behavior was established for n-alkyltitanium compounds [21,22,77,78]. For example, quenching ethyllithium *25* with chlorotitanium triisopropoxide *3* affords *9*, which reacts in situ with a 1:1 mixture of *19* and *20* to afford essentially only the aldehyde adduct *23* (Equation 9) [77]. *25* itself reacts at 0 °C almost statistically!

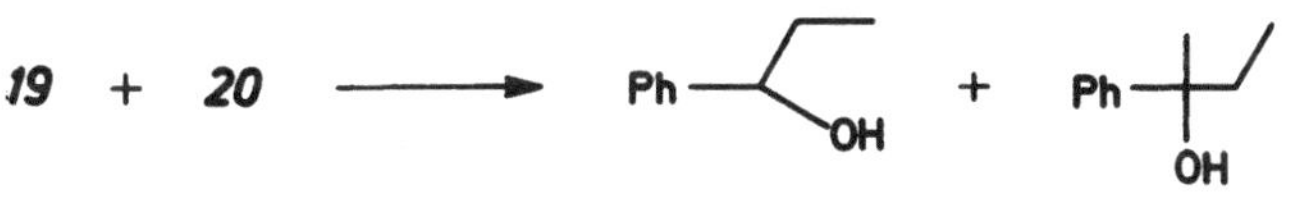

| | 23 | 24 |
|---|---|---|
| $CH_3CH_2Li$ 25 | 51 | 49 |
| $CH_2CH_2Ti(OCHMe_2)_3$ 9 | >99 | <1 |

(9)

Essentially complete chemoselectivity (>99%) is also observed in reaction of *6*, *9* and *10* with mixtures of aliphatic aldehydes (e.g., hexanal) and ketones (e.g., 2-heptanone) [77]. The results show that the difference in activation energy ($\Delta\Delta G^{\neq}$) for aldehyde and ketone addition must amount to several kcal or more. In fact, preliminary kinetic studies involving *26* and *27* reveal a relative rate of $k_{rel} = k_{26}/k_{27} = 223$ [79].

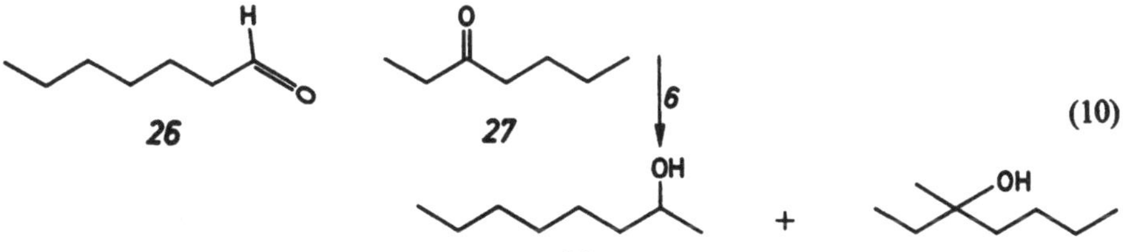

26 27 28 29

(10)

Synthetically, there are some limitations to the use of organotitanium reagents. Attempts to quench secondary alkylmagnesium reagents with *3*[3] followed by the addition of benzaldehyde *19* did not result in high yields of Grignard-type adducts (<50%) [77]. Instead, reduction and reductive dimerization of *19* dominated, a result of undesirable transitionmetal-like behavior (see Sect. B). An exception is cyclopropyltitanium triisopropoxide *11* (formed from cyclopropyllithium and *3*), which adds in situ to benzaldehyde (83%) [77]. Decomposition such as β-hydride elimination is kinetically less favorable due to ring strain in cyclopropene. Certain vinyltitanium compounds

---

[3] Secondary Grignard compounds isomerize to primary isomers in the presence of catalytic amounts of $TiCl_4$ [8h].

undergo rapid oxidative coupling [80]. In these cases the zirconium analogs (although more expensive [81]) are better suited [80]. For example, quenching tert-butyl- or 1-cyclohexenyllithium with chlorozirconium tributoxide results in zirconium reagents which are highly aldehyde-selective; unfortunately, the yields of adducts or amounts of side products were not listed [80]. Another limitation pertains to the use of alkyltitanium triamide derivatives. In situ generation of *30* from methyllithium and chloro- or bromotitanium(tris)diethylamide) *15* followed by addition of benzaldehyde *19* (0 °C, 0.5 h) and aqueous workup did not result in the Grignard-type adduct *21*. Instead, about 35% *31* was isolated in addition to recovered *19* and an oligomeric residue [77]. Longer reaction periods did not lead to the complete consumption of *19*, so that *32* may well be involved, which reverts back to *19* upon hydrolysis.

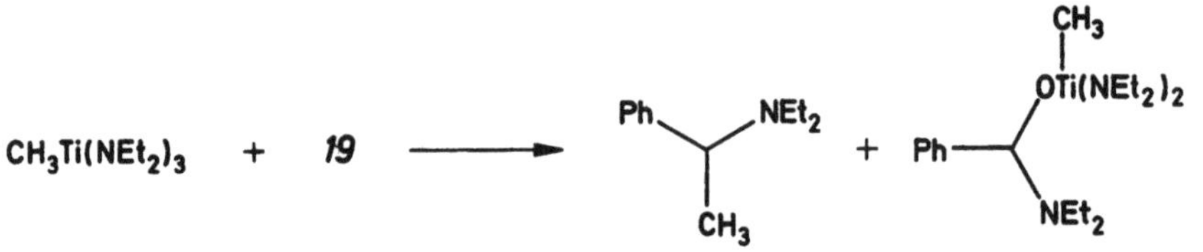


(11)

*30* *31* *32*

The synthetic drawbacks of amino-substituted reagents are *not* observed when using resonance stabilized "carbanions" such as α-deprotonated nitriles [77] and carbonyl compounds [25], or allyl- [21] and benzylmagnesium bromide [77], the corresponding very reactive tris-aminotitanium reagents undergoing rapid (−78 °C to −20 °C, 0.5 h) Grignard-type addition (80–95% yields).

A number of resonance stabilized or heteroatom substituted carbanions have been converted into titanium analogs for the purpose of testing chemoselectivity, e.g., Equation 12 [21,77]. Generally, an in situ reaction mode was chosen. In all cases a high degree of aldehyde-preference was noted.

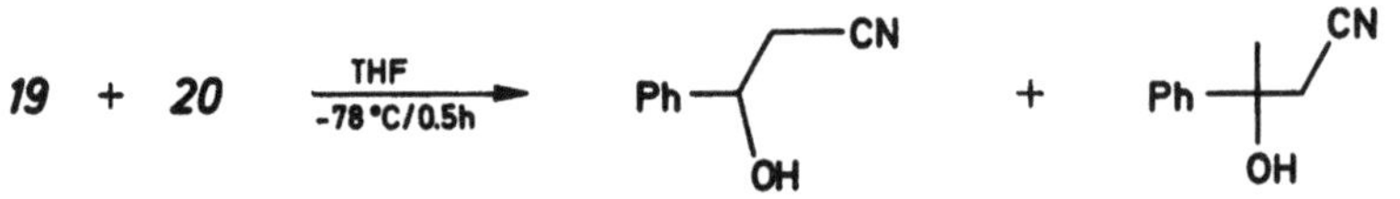


| | *33* | *34* |
|---|---|---|
| $NCCH_2Li$ | 64 | 36 |
| $NCCH_2Ti(OCHMe_2)_3$ | >97 | <3 |

(12)

Similar selectivities (>98%) have been reported for *35–38*, although the actual yield of addition product is not always satisfactory [78].

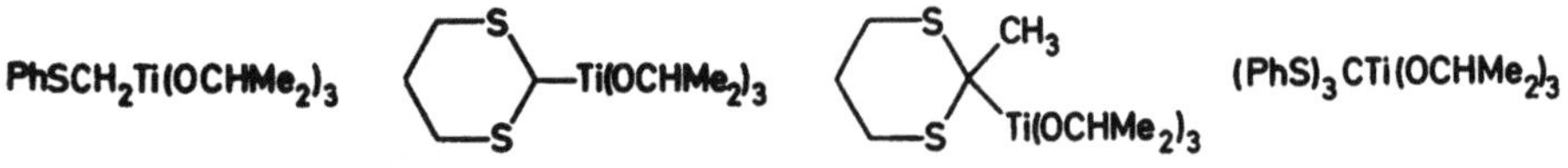


*35* *36* *37* *38*

## C.III Types of Additional Functional Groups Tolerated

Aldehydes *39a–g* — *54* react smoothly with methyltitanium triisopropoxide *6* (the numbers in parentheses refer to yields of isolated adducts). Generally, a 10–15% excess of titanium reagent is used, except in compounds containing HO-groups (100% excess). In case of α,β-unsaturated carbonyl compounds, the 1,2-addition mode is observed [22, 76, 82].

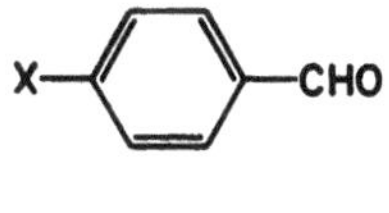

*39*

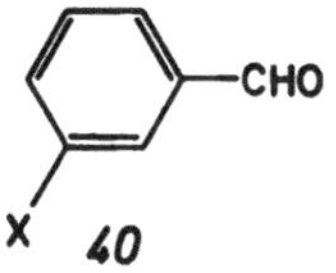

*40*

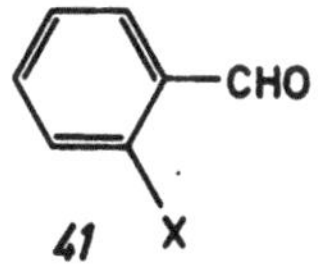

*41*

*39*:
*a* X = $NO_2$ (95%)[78,82]
*b* X = CN (85%)[82]
*c* X = $CO_2H$ (79%)[82]
*d* X = $OCH_3$ (86%)[82]
*e* X = F (80%)[82]
*f* X = Cl (94%)[82]
*g* X = Br (96%)[82]
*h* X = I —

*40*:
*a* X = $NO_2$ (82%)[82]
*b* X = Br (98%)[82]

*41*:
*a* X = $NO_2$ (87–92%)[78,82]
*b* X = Br (94%)[82]

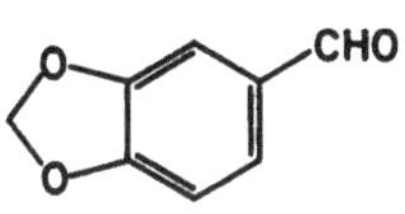

*42* (90%)[82]

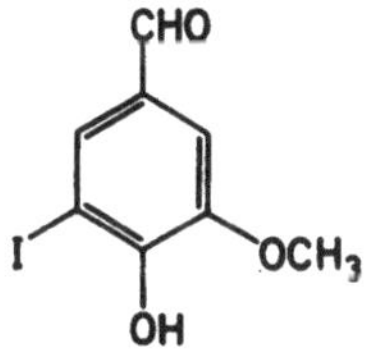

*43* (41%)[82]

*44* (83%)[82]

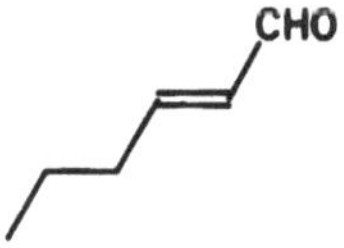

*45* (94%)[82]

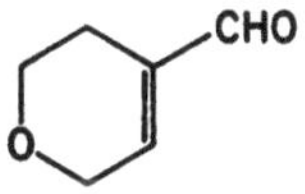

*46* (84%)[82]

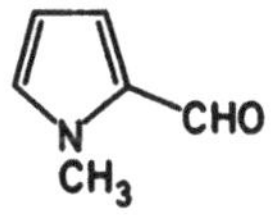

*47* (83%)[82]

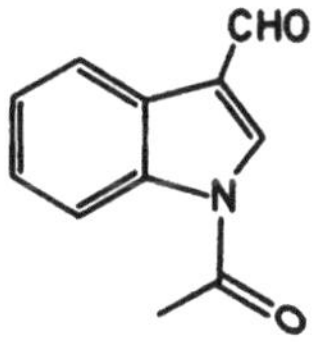

*48* (70%)[82]

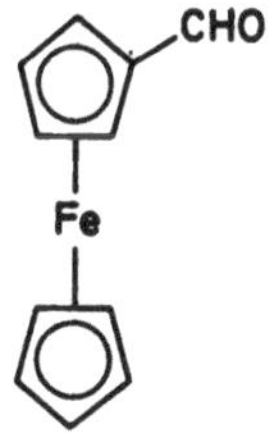

*49* (92%)[82]

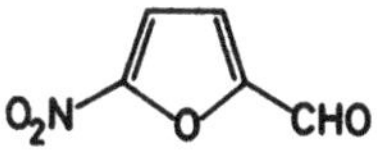

*50* (79%)[82]

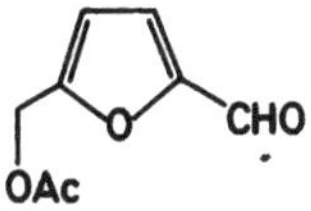

*51* (81%)[82]

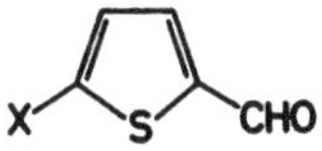

*52*

*a* X = H (88%)[82]
*b* X = Br (89%)[82]

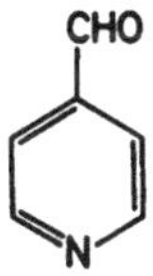

*53* (76%)[82]

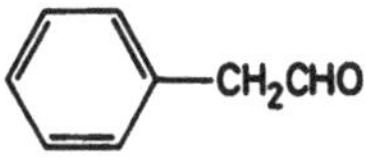

*54* (74%)[82]

Furthermore, the n-butyl derivative *10* adds in situ to *39b* (97% yield)[78], *39g* (75%)[82], *39h* (82%)[78], *41b* (92%)[78]. The addition of phenyltitanium triisopropoxide *5* to *39a* proceeds smoothly (94%)[78].

Although many of the aldehydes also react with lithium and magnesium reagents, the above studies were necessary in order to show that titanium, being a transition metal, reacts at least as well. Indeed, better performance was noted in many cases. A particularly dramatic example is the reaction of n-butyllithium with *39h*, which *fails* to afford any of the corresponding adduct due to rapid I/Li exchange and other undesired side reactions[78]. Also, *48* and *51*, which contain several functional groups, were checked with methyllithium and methylmagnesium bromide, the highest yield being less than 45%[82]. Finally, *6* reacts selectively with aldehydes in the presence of added epoxides[66, 76], esters[66, 76] and thio-esters[76].

A few chemoselectivity studies were also carried out with ketones containing additional functional groups. Various organotitanium reagents react with levulinic acid ethylester *55* solely at the keto-function, but subsequent lactonization may occur[72, 83]. For example, *6* affords the adduct *56* which undergoes cyclization to form *57* (~90%) prior to aqueous workup[83]. Although a six-membered lactone is formed analogously from the corresponding keto-ester, it remains to be seen whether titanium reagents can be used to synthesize larger lactones[77].

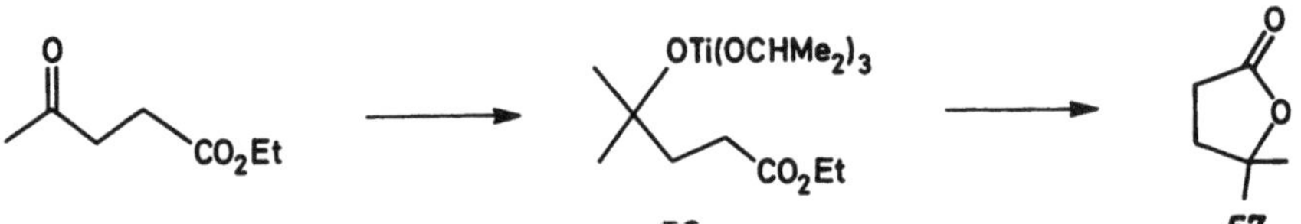


55 56 57 (13)

The methylation of pregnenolone acetate *58* to form *59* occurs chemoselectively with *6*[83]. The *zirconium* reagent *61*, prepared by quenching methyllithium with *60*, reacts analogously[83], so that no advantages are apparent in this case.

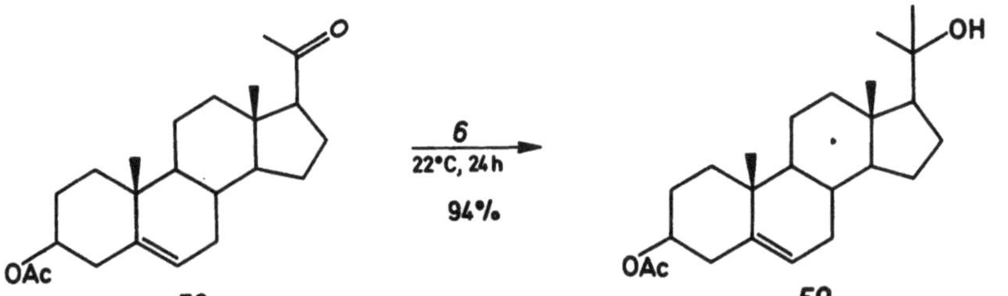


58 59 (14)

$$ClZr(OPr)_3 \xrightarrow{CH_3Li} CH_3Zr(OPr)_3 \quad (15)$$

60 61

Cyano or chloromethyl groups also do not interfere with carbonyl addition. Dimethyltitanium diisopropoxide *63*[21], easily accessible from *62*[84], reacts chemoselectively with *64* to form *65*[4] (95%)[82]. *63* is considerably more reactive than

[4] A single diastereomer resulting from preferential equatorial attack is formed. Stereoselectivities are discussed in detail in Section D.II.

the monomethyl derivative *6* (see Sect. C.VI), and in the present case the amount of reagent chosen was such that only one methyl group was transferred. *66* transforms readily to *67* (Equation 18).

$$1 + 2 \longrightarrow \underset{62}{Cl_2Ti(OCHMe_2)_2} \xrightarrow{CH_3Li} \underset{63}{(CH_3)_2Ti(OCHMe_2)_2} \quad (16)$$

64 65 (17)

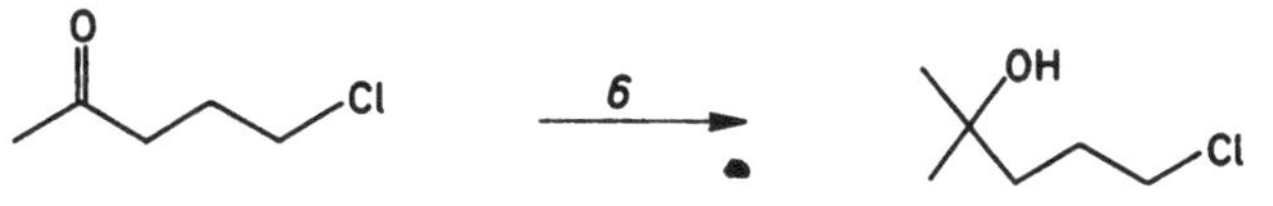


66 67 (18)

A final point regards the incompatibility of *aliphatic* nitro groups with organotitanium reagents, claimed in the literature [80]. It was reported that neither *68* nor *70* are methylated by methyltitanium triisopropoxide *6* [80]. Our own investigations in this area showed that the reaction of *68* with *6* is in fact sluggish, leading to about 20% of *69* as well as unidentified sideproducts after 2 days at rooms temperature. However, *63* is an excellent methylating agent as shown in Equation 19. Compounds *68* and *70* have been methylated using zirconium reagents [80].

68 69 (19)

*6* (22 °C/ 2d) ~20%

*63* (22 °C/ 2h) 75–80%

70 71 (20)

Hesse's report [85] that *72* is smoothly methylated by *6* (85% *73*) also demonstrates that incompatibility of aliphatic nitro groups with organotitanium reagents is not general. In this case a more reactive aldehyde function is involved. As already stressed, ketones react much more slowly than aldehydes, so that side reactions involving the aliphatic nitro group (e.g., deprotonation of *68*) begin to compete. In such cases, either more reactive titanium reagents, e.g., *63*, must be used (albeit with loss of one active methyl group), or the corresponding mono-methyl *zirconium* compounds. Similar advantages of certain zirconium reagents relative to titanium analogs pertain to the alkylation of enolizable ketones (see Sect. C.V). However, they do not discriminate between aldehydes and ketones as well as the titanium counterparts [80].

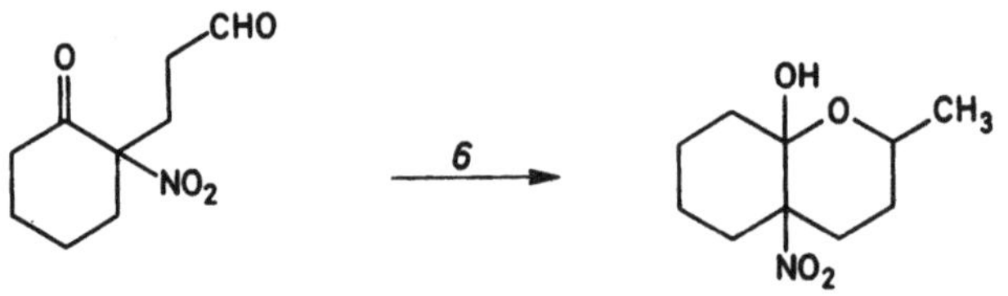


*72* *73* (21)

Why are organotitanium reagents more chemoselective than the lithium and magnesium counterparts? It is clear from the above presentation that the rate of reaction of classical carbanions is considerably higher than those of the titanated species. Generally, more reactive species are less selective. However, this does not really answer the above question, since the phenomenon of different rates remains unclear. Apart from steric factors, we believe it has to do with the different *polarity* of the carbon-metal bond. Whereas C—Li bonds are highly polar (and ionic in certain cases) [57, 86, 87] the C—Ti analogs appear to be considerably less so (Sect. B).

## C.IV Discrimination between Compounds belonging to the same Funtional Group

If a molecule contains several carbonyl groups belonging to the *same* functional group (e. g., two keto groups), the problem of site-selectivity arises. Fundamental information can be derived not only by studying the product distribution in reactions of such compounds, but also by performing intermolecular competition experiments using two different compounds having the same type of functional group (e.g., two different ketones). Encouraged by the results of the above chemoselectivity studies, we anticipated a certain degree of selectivity.

The first system to be studied involved the addition of methyltitanium triisopropoxide *6* (one part) to a 1:1 mixture of hexanal *74* and 2-ethylbutanal *75* [21]. The product ratio *76*:*77* turned out to 92:8 (95% conversion), showing that reactions of organotitanium reagents are very sensitive to the steric environment around the reaction center. It is interesting to note that in Grignard reactions a reversed mode of addition is required, i.e., the aldehyde is added to the very basic organometallic reagent, otherwise rapid aldolization sets in. For competition experiments this is unsuitable [21].

In the ketone series, the result of reacting *6* (one part) with a mixture of cyclohexanone *78* (one part) and 4-heptanone *79* (one part) was rather unexpected. The information can be derived not only by studying the product distribution in reactions (92% conversion) [82]! In contrast, methyllithium affords a 2:3 mixture of *80* and *81*.

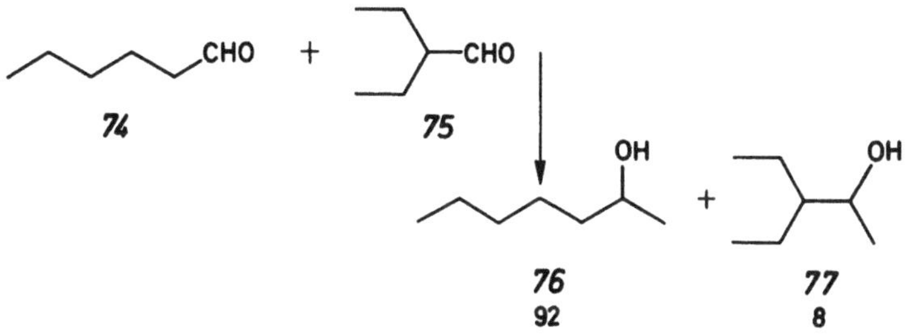


(22)

Neglecting minor electronic differences, the preferred addition to *78* can be ascribed to steric factors.

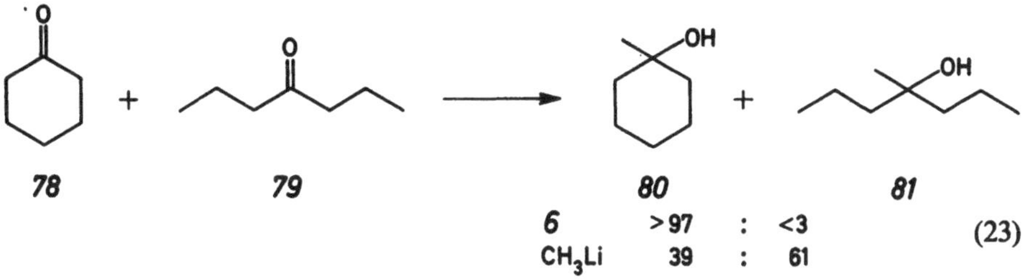


(23)

Although somewhat less selective, the reaction of *6* with 4-heptanone *79* and 2-heptanone *82* to form a 15:85 mixture of *81* and *83* (90% conversion, Equation 24) is also impressive, particularly in view of the fact that methyllithium reacts statistically [82].

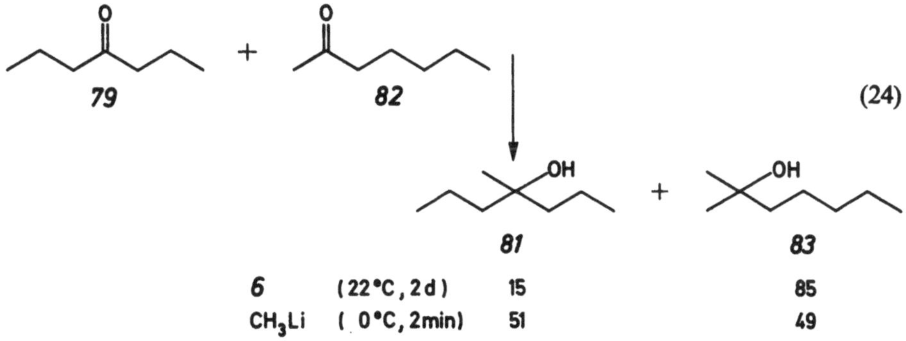


(24)

Also, we have tested pairs of ketones (and aldehydes) in which one of the members is *electronically* different. An example is depicted in Equation 25. The carbonyl C-atom in the unsaturated ketone is less electrophilic due to resonance and therefore less reactive, inspite of the fact that arguments based on steric consideration predict the opposite.

A final example is shown in Equation 26 [77]. In conclusion, the present results demonstrate that organotitanium reagents are *highly sensitive to steric and electronic effects* in the carbonyl substrate. This means that application in more complex systems, e.g., natural product syntheses, is likely to be successful.

78 + 84 ⟶ 80, 85 (25)

| | 80 | 85 |
|---|---|---|
| *6* | 94 | 6 |
| $CH_3Li$ | 80 | 20 |

*1a* + *86* ⟶ *21* + *87* (26)

| | 21 | 87 |
|---|---|---|
| *6* | 13 | 87 |
| $CH_3Li$ | 1 | 1 |

## C.V Addition to Enolizable Ketones

Enolizable ketones often afford low yields of addition products with Grignard [64a] and alkyllithium [64b] reagents due to competing deprotonation. For example, *88* [21] reacts with methyllithium to yield less than 55% of *89*[5]. Whereas *6* is not much better, dimethyltitanium diisopropoxide *63* affects smooth methylation (90%) [21]. A case in which titanium is *less* suited than lithium involves methylation of α-tetralone *90*. It was reported that *6* yields not more than 50% of the addition product *91* [80,83], while *63* [83] or zirconium reagents [80,83] lead to 90% conversion. Thus, the zirconium compound is clearly better than *6*, but the conclusion that this is due to the higher basicity of titanium compounds [80] is not completely satisfactory because it focuses only on *one* of two competing processes (deprotonation vs. addition). Indeed, addition of methyllithium, which is certainly more basic than *6*, poses no problems [77] (Equation 28)!

*88* ⟶ *89* (27)

| | 89 |
|---|---|
| $CH_3Li$ | <55% |
| $(CH_3)_2Ti(OCHMe_2)_2$ *63* | >90% |

[5] Diastereoselectivities associated with reactions of *88* will be discussed in Section D.II.

*90*      *91*

| | | |
|---|---|---|
| $CH_3Li$ | 85% | (28) |
| *6* | 50% | |
| *63* | 90% | |
| $CH_3Zr(OR)_3$ | 90% | |

Grignard-type addition to the sterically hindered ketone *92* is a challenge because methylmagnesium iodide [88a)] and methyllithium [88b)] lead to >90% enolization. Unfortunately, *6* and *63* as well as the zirconium reagent *61* also fail. Based on the observation that reagents containing two methyl groups are more reactive and less prone to induce enolization [21)], we decided to test tetramethyltitanium *14* and the zirconium analog *95* [89)], both of which are known to be stable below —20 °C. Indeed, they proved to be "super" methylating agents at low temperatures [83)]. *14* affords about 20% *93*, 35% *94* and 45% starting material. *95* yields 45% 93 and 55% 92, but no addition/elimination product *94*. Thus, *tetramethylzirconium 95 is a highly reactive reagent of low basicity* [83)]. Both reagents permethylate CH-acidic β-keto-esters [77)]. We believe that successive substitution of alkoxy ligands by methyl groups at titanium increase reactivity because the stabilizing effect of π-donation is thereby reduced (Sect. B.III).

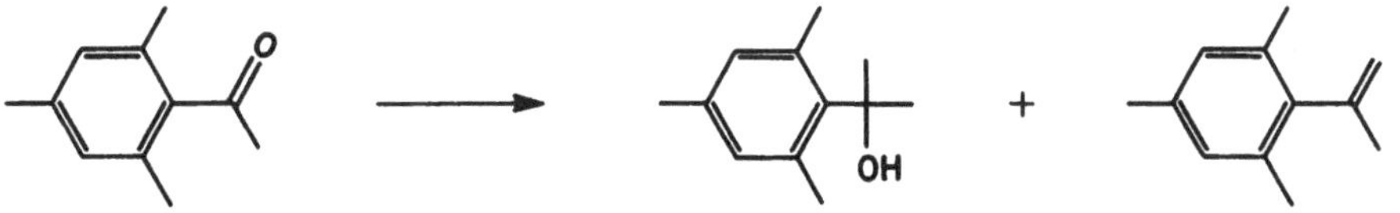


| | *92* | *93* | *94* (29) |
|---|---|---|---|
| $(CH_3)_4Ti$ *14* | | 20% | 35% |
| $(CH_3)_4Zr$ *95* | | 45% | 0% |

## C.VI Relative Rates of Carbonyl Addition

Although detailed kinetic studies are not yet available, qualitative observations regarding the addition of $RTi(OCHMe_2)_3$ to carbonyl compounds point to the following *approximate* reactivity scale: [77)]

> Allyl > cyanomethyl ~ benzyl ≫ methyl > phenyl ~
> cyclopropyl ~ ethyl ~ n-butyl ≫ o-fluorophenyl ~
> trimethylsilylmethyl ≫ per-fluorophenyl

Increasing the number of alkyl groups at titanium increases the rate of addition dramatically:

$$CH_3Ti(OR)_3 < (CH_3)_2Ti(OR)_2 < (CH_3)_3TiOR < (CH_3)_4Ti$$

Within the series $CH_3Ti(OR)_3$, the rate increases slightly with increasing size of the alkoxy groups [78,83]. Sterically, this is opposite to what one would expect. We believe that the rate differences are due to different degrees of aggregation (see Sect. B.I). Compounds having alkoxy groups smaller than isopropoxy are dimeric in which titanium is penta-coordinate, and this seems to decrease the rate of addition to carbonyl compounds.

A final point regards the effect of lithium salts. Methyltitanium triisopropoxide *6*, prepared from methyllithium and *3*, reacts in situ with carbonyl compounds lightly slower than the distilled compound. However, this has no effect on chemoselectivity[6]. The only exception noted so far concerns allyl derivatives, as discussed in the following Section C.VII.

## C.VII Controlled Reversal of Chemoselectivity using Titanium Ate Complexes

All of the titanium reagents discussed so far are highly aldehyde-selective. Allyl metal compounds are known to be considerably more reactive than the corresponding n-alkyl analogs. Thus, in case of allyl titanium reagents, we anticipated a lower degree of aldehyde selectivity. Indeed, allyltitanium triisopropoxide *97*, made by quenching allylmagnesium chloride *96* with *3*, reacts with heptanal *26* and 2-heptanone *82* to form an 84:16 mixture of *100* and *101*, respectively [90]. The Grignard *96* and lithium compounds *107* are slightly ketone-selective, while the zirconium reagent *98* (in situ from *96* and *60*), is essentially non-selective (Table 3). In hope of optimizing chemoselectivity, we prepared the titanium ate complex *99*[7] by quenching *96* with titanium tetraisopropoxide *2*. Addition to a 1:1 mixture of *26* and *82* resulted in smooth product formation (>90% conversion), in this case the *100*:*101* ratio being 98:2 (Table 3).

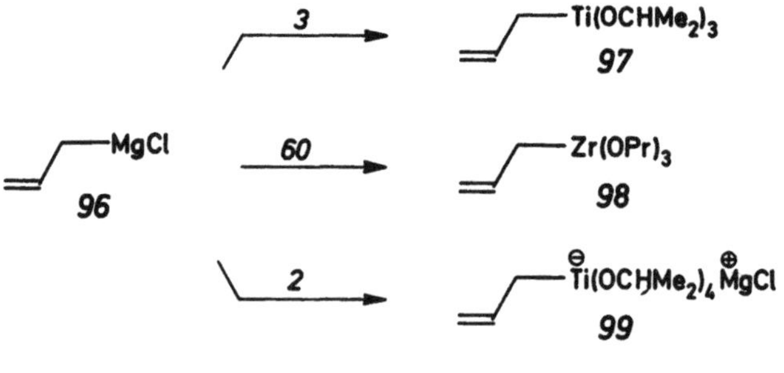


(30)

(31)

(32)

[6] The chemoselectivity studies involving *6* (Sect. C.II) were performed with distilled reagent. In situ procedures were used for ethyl and n-butyl derivatives (*9* and *10* plus Li salts), the reagent being completely aldehydeselective.

[7] The precise structure is currently unknown.

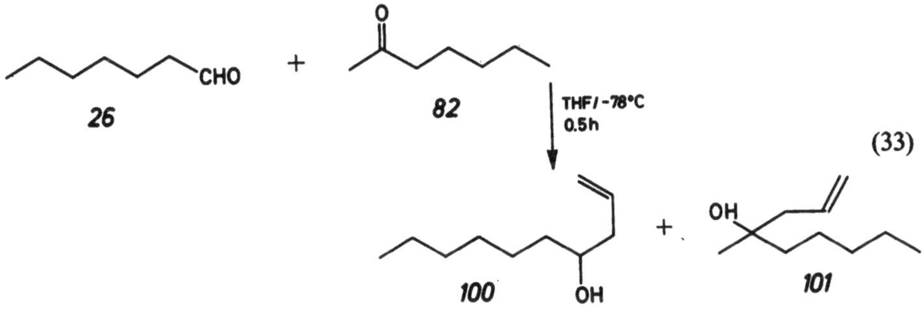


(33)

Upon testing the effect of other ligands at titanium, we came across some unexpected results [90]. The amino derivative *102*, prepared from *96* and *15*, reacts in situ preferentially with *ketones*. In case of *26* and *82* the *100*:*101* ratio is 22:78 (Table 3). The in situ reaction of the related compound *104* leads to a 13:87 ratio. By using the ate complexes *106*, *108* and *110*, the ketone-selectivity turned out to be 4:96, 2:98 and 14:86, respectively!

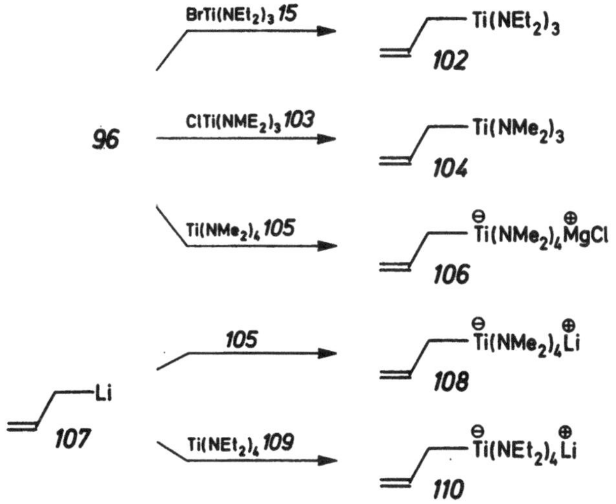


(34)

(35)

(36)

(37)

(38)

Although a few observations relevant to the mechanism have been made, it is difficult to offer a clear explanation at this time. Initial control experiments indicate that the reactions are kinetically controlled, i.e., they are irreversible under the conditions used. Furthermore, a salt effect is involved, because the distilled aminotitanium compound *102* is essentially non-selective (Table 3). Upon adding $MgX_2$ salts, ketone selectivity increases to ~70%.

Intrigued by the above results, we tested the generality of the above phenomena. Indeed, upon using benzaldehyde *19* and acetophenone *20* (eq. 39), very similar results were obtained (Table 4).

Since substituted allyl titanium derivatives (e.g., crotyl) with amino ligands are sterically more demanding than the parent compound, they should display lower degrees of ketone-selectivity. However, this turned out *not* to be the case. The

**Table 3.** Chemoselective Addition of Allyltitanium Reagents to Aliphatic Carbonyl Compounds [90]

| Reagent | Ratio *100*:*101* |
|---|---|
| *96* | 29:71 |
| *97* (in situ) | 86:14 |
| *99* | 98: 2 |
| *102* (in situ) | 22:78 |
| *102* (distilled) | 46:54 |
| *104* (in situ) | 13:87 |
| *106* | 4:96 |
| *107* | 31:69 |
| *108* | 2:98 |
| *110* | 14:86 |

**Table 4.** Chemoselective Addition of Allyltitanium Reagents to Aromatic Carbonyl Compounds [90]

| Reagent | Ratio *111*:*112* |
|---|---|
| *96* | 38:62 |
| *97* (in situ) | 84:16 |
| *99* | 98: 2 |
| *102* (in situ) | 50:50 |
| *104* (in situ) | 12:88 |
| *106* | < 1:99 |
| *108* | 9:91 |

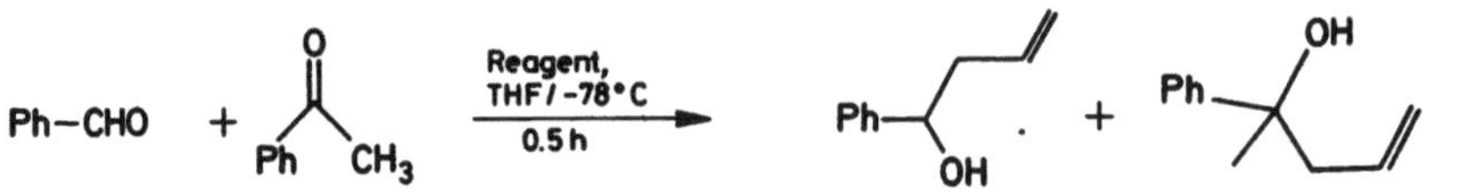


*19* *20* *111* *112* (39)

results of competition experiments using aliphatic carbonyl compounds (Equation 40) as well as the aromatic pair (Equation 41) in combination with crotyl metal compounds *113–116* are shown in Table 5. It should be noted that in all cases α-selectivity is observed, which means that threo/erythro diastereomers result. The latter aspect will be discussed separately (Sect. D.III.3)

In summary, it is possible to reverse chemoselectivity in a predictably manner simply by varying the nature of the ligands at titanium. Since the olefinic moiety in the products can be manipulated chemically (e.g., reduction, oxidation), the possibility of chemoselective delivery of various functional groups emerges.

The ate complex *99* also reacts more rapidly with ketones than with esters, as exemplified by the smooth reaction with levulinic acid ethylester *55* (>90% con-

**Table 5.** Chemoselective Addition of Crotyltitanium Reagents to Carbonyl Compounds [90]

| Reagent | Ratio *117*:*118* | Ratio *119*:*120* |
|---|---|---|
| *113* | 26:74 | — |
| *114* | 98: 2 | — |
| *115* (in situ) | 18:82 | — |
| *116* | 1:99 | — |
| *113* | — | 34:66 |
| *114* | — | 99: 1 |
| *115* | — | 11:89 |
| *116* | — | 2:98 |

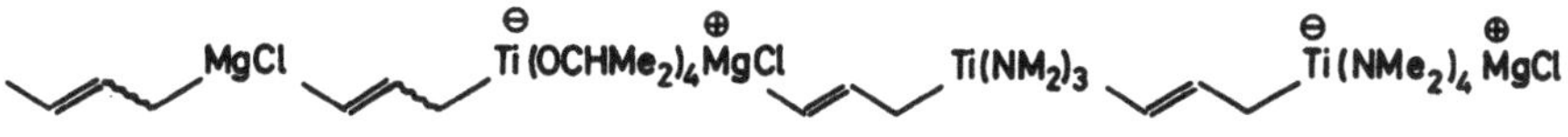


*113* *114* *115* *116*

*26* + *82* 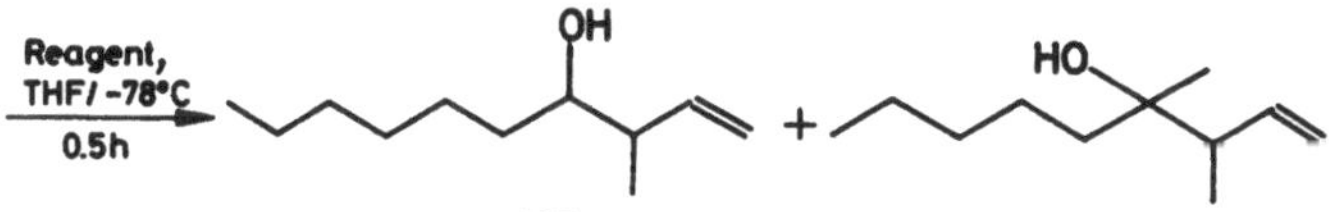


*117* *118*

(40)

*19* + *20* 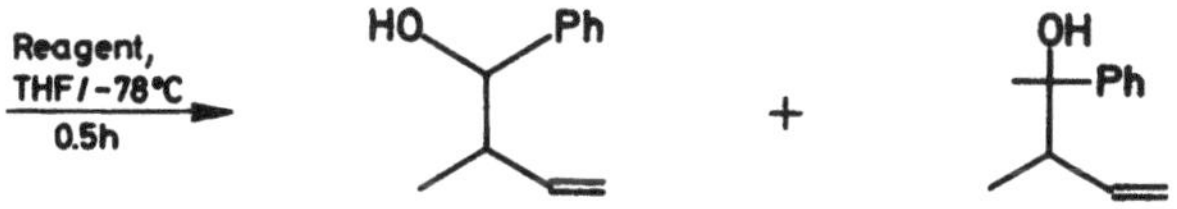


*119* *120*

(41)

version, 80% of isolated *121*) [90]. Allyllithium or-magnesium are not chemoselective and afford mixtures.

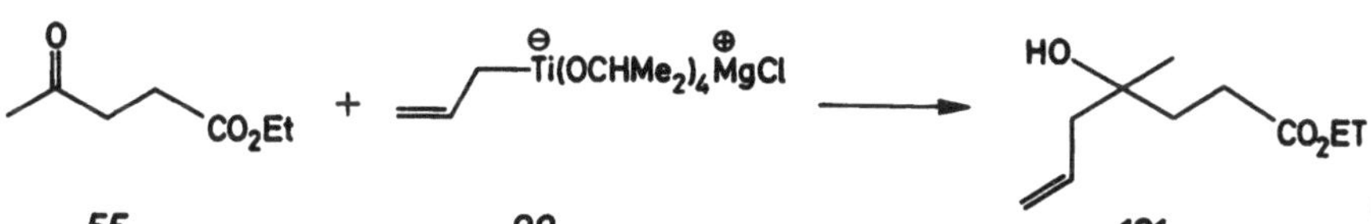


*55* *99* *121*

(42)

It is interesting to compare the above results with the reactions of other allyl metal reagents. Whereas allyllithium and allylmagnesium are slightly ketone-selective (Tables 3–5), allylboranes [91] as well as certain allylchromium derivatives [92] react faster with aldehydes. Allyltin reagents are also aldehyde-selective (Equation 43)[93].

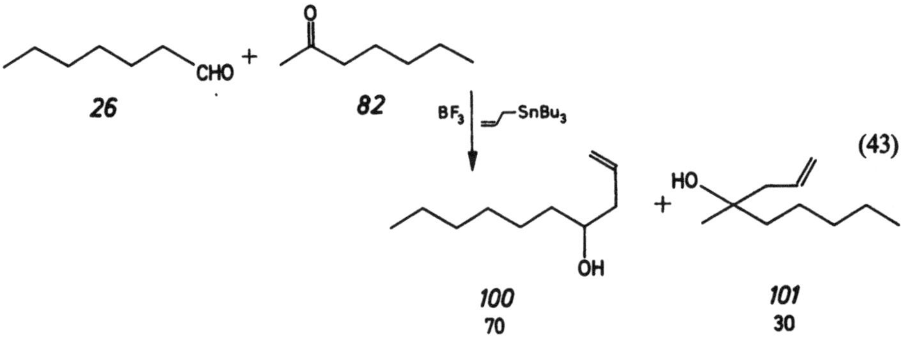


(43)

The unique phenomena observed for allyltitanium ate complexes appear to be restricted to allyl groups. Reversal of chemoselectivity in methylation reactions has not been observed to date. Ate complexes of the type *122* and *123* react much faster with aldehydes than with ketones[90], but the reactions are not as clean as with the allyl derivatives. Aromatic nitro and cyano groups are tolerated. In certain situations reversal of *diastereoselectivity* is possible using *123* (Sect. D.II). Ate complexes prepared from poly-titanates[17] are also aldehyde selective.[77]

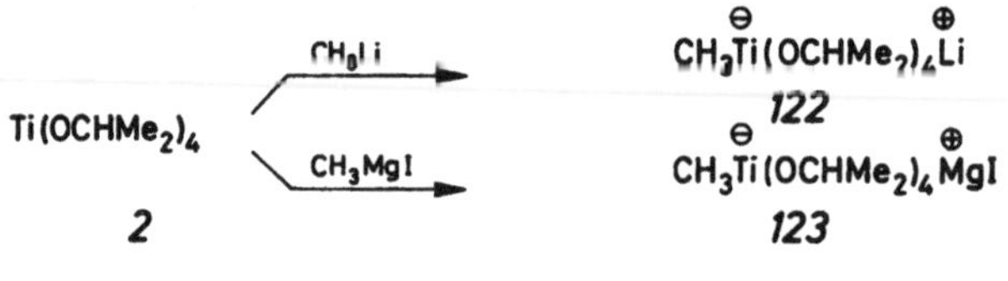


(44)

(45)

# D. Diastereoselectivity in Reactions of Organotitanium Reagents with Carbonyl Compounds

## D.I Asymmetric Induction in Acyclic Systems

In a series of important papers, Cram reported 1,2- and 1,3-asymmetric induction in the addition of Grignard and alkyllithium reagents to chiral aldehydes and ketones[94]. The results led to several models with high degrees of predictive power[94,95]. A widely quoted example is the addition of Grignard and alkyllithium reagents to 2-phenyl-propanal *124*[8] to form the diastereomers *125* and *126*, generally in a ratio of about 2:1[94,95].

---

[8] Here as well as in the following examples only one enantiomer is shown arbitrarily, although a racemate was used.

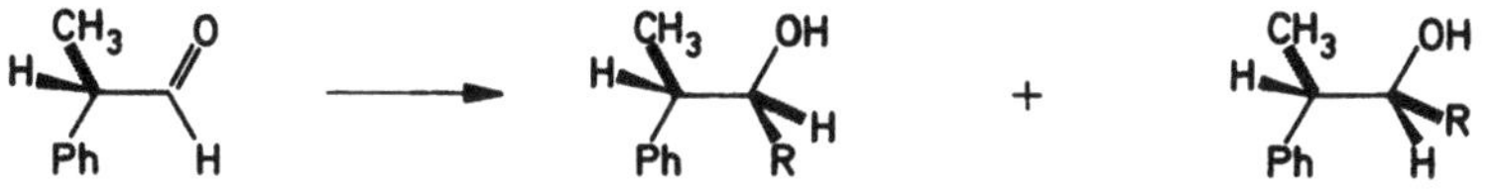

124 125 126 (46)

In the first reported example of asymmetric induction using organotitanium reagents, methyltitanium triisopropoxide *6* was reacted with *124* (0 °C/2 h, THF) [72]. The ratio of Cram to anti-Cram product *125*:*126* turned out to be 88:12 (Table 6) which is higher than that observed for $CH_3MgX$ (66:34 = or $CH_3Li$ (65:35) [94,95]. Variations due to temperature or solvent effects are small. The tri-phenoxy derivative *127* (Table 6) displays the highest degree of selectivity. The zirconium reagents *61*

**Table 6.** 1,2 Asymmetric Induction in the Addition to *124*

| Reagent | R in *125* and *126* | *125*:*126* | Ref. |
|---|---|---|---|
| $CH_3Ti(OCHMe_2)_3$ | Methyl | 88:12 | 22,72) |
| $CH_3Ti(OPh)_3$[a] *127* | Methyl | 93: 7 | 77) |
| $CH_3TiCl_3$ *17* | Methyl | 81:19 | 72) |
| $CH_3Zr(OPr)_3$ *61* | Methyl | 90:10 | 83) |
| $CpTi(CH_3)_3$[b] *128* | Methyl | 86:14 | 77) |
| $CH_3MgBr$ | Methyl | 66:34 | 94,95) |
| $CH_3Li$ | Methyl | 65:35 | 94,95) |
| n-$C_4H_9Ti(OCHMe_2)_3$ *10* | n-Butyl | 89:11 | 83) |
| n-$C_4H_9MgBr$ | n-Butyl | 66:34 | 77) |
| $CH_2{=}CHCH_2Ti(NEt_2)_3$ *102* | Allyl | 93: 7 | 21) |
| $CH_2{=}CHCH_2MgCl$ *96* | Allyl | 65:35 | 77) |
| $(C_6H_5)_2TiCl_2$ *129* | Phenyl | 80:20 | 72) |
| PhMgX | Phenyl | 80:20 | 94,95) |

[a] Prepared from chlorotitanium triphenoxide and $CH_3Li$; in situ reaction with *124* in ether at −50 °C

[b] Prepared from $CpTiCl_3$ *18* and three equivalents of $CH_3Li$; addition of *128* to *124* was carried out in THF, the relative amounts chosen so that only one methyl group was transfered.

behaves much like the titanium compound. n-Butyl- and allyltitanium compounds (*10* and *102*) also display greater distereoselectivity than the corresponding Grignard reagents, whereas selectivities of phenyl derivatives (e.g., *129*, Table 6) do not vary much as a function of the metal. The results can be explained by Cram's original open-chain model or by more recent theories [95]. It is interesting to note that organozinc and cadmium reagents in the presence of activating salts are even *less* diastereoselective than $CH_3MgX$ or $CH_3Li$ [95c].

As already shown in Section B.I, certain organotitanium reagents readily form isolable, octahedral 1:2 adducts with such donor molecules as THF, glyme, thioethers, amines and diamines [1,19] (Equation 47). In case of methyltitanium trichloride *17*, structural data show the methyl group to occupy the equatorial position [96]. In order to test whether such molecules undergo stereoselective addition to aldehydes (Equation 47), we reacted *134*, *135* and *136* (prepared from TMEDA, glyme and THF, respectively) with 2-phenylpropanal *124* [97]. The *125*:*126* ratios of 80:20, 82:18 und 85:15 show that the Cram product is preferred in all cases (conversion ~80%). The THF-adduct *136* displays the highest degree of diastereoselectivity. The mechanism of addition involves ligand exchange *131* → *132* in an octahedral complex in which the methyl group is displaced by an alkoxy ligand prior to aqueous workup (Equation 47). This may occur via a heptacoordinate intermediate or transition state ($S_N2$), or by prior dissociation of chloride to form a short lived pentacoordinate species ($S_N1$)[9]. The former seems more likely, since related hexa-coordinate species isomerize (diastereoisomerization) via pseudorotation and not by dissociation/recombination [96].

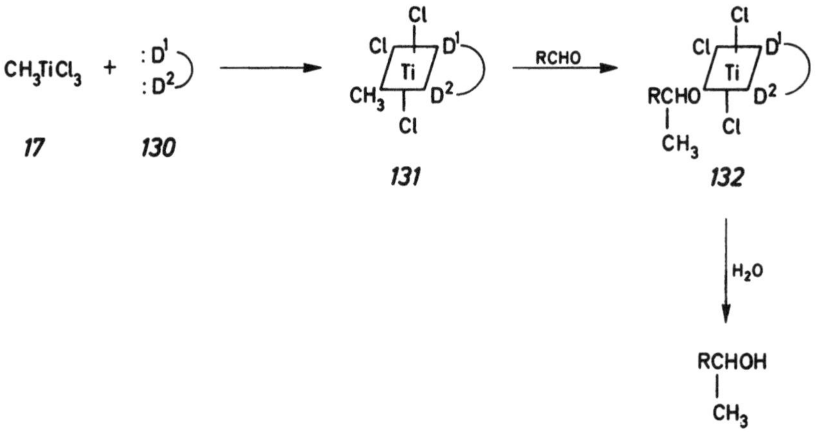


*17* *130* *131* *132* *133* (47)

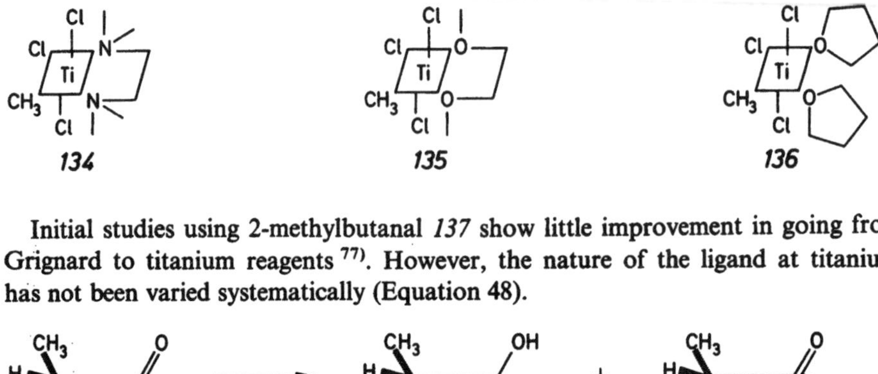


*134* *135* *136*

Initial studies using 2-methylbutanal *137* show little improvement in going from Grignard to titanium reagents [77]. However, the nature of the ligand at titanium has not been varied systematically (Equation 48).

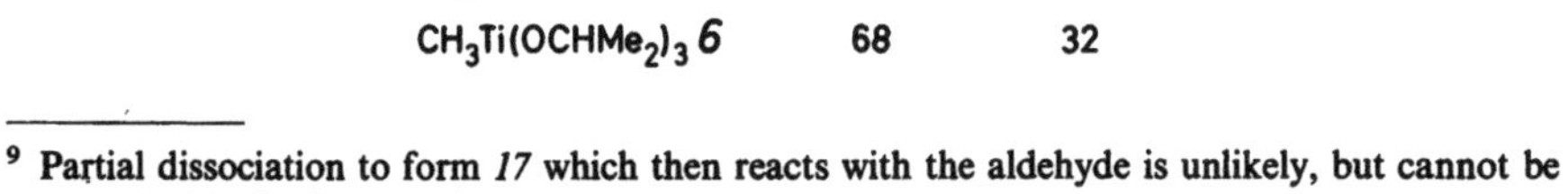


*137* *138* *139* (48)

| | 138 | 139 |
|---|---|---|
| $CH_3MgBr$ | 60 | 40 |
| $CH_3Ti(OCHMe_2)_3$ *6* | 68 | 32 |

---

[9] Partial dissociation to form *17* which then reacts with the aldehyde is unlikely, but cannot be excluded with absolute certainty at present.

We have also studied more complex systems. Titanium reagents are well suited for the stereoselective construction of steroidal side chains. For example, pregnenolone acetate *58* reacts with allyltitanium triisopropoxide *97* or $D_3$-methyltitanium triisopropoxide *6a* to form mixtures of *140* and *141* in ratios of >90:10 and 96:4, respectively [83]. The corresponding Grignard reagents are less selective (83:17 [77] and 88:12 [98], respectively). The results can be interpreted by assuming selective attack at the si face of the carbonyl group in the so called "D-ring conformation" [98,99]. Surprisingly, the zirconium reagent *61a* reacts stereo-randomly [83].

A clear limitation of titanium chemistry has to do with the observation that such sterically hindered ketones as *58* do not react well with the less reactive n-alkyltitanium reagents [77]. Of course, this does not apply to sterically non-hindered steroidal $C^{22}$-aldehydes, which react smoothly and with high degrees of asymmetric induction [77].

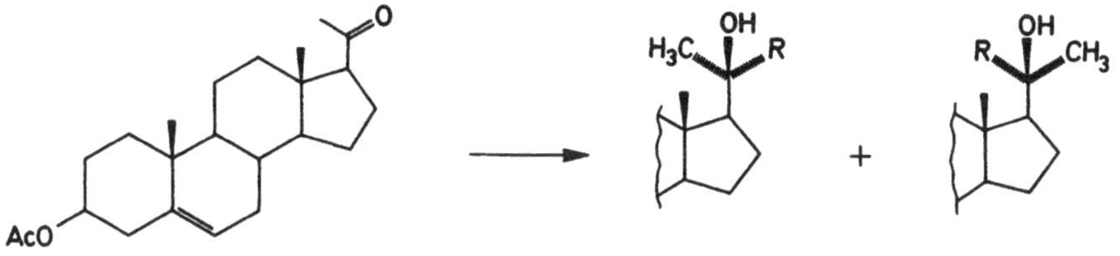


| | *58* | *140* | *141* | (49) |
|---|---|---|---|---|
| $CH_2{=}CH_2CH_2Ti(OCHMe_2)_3$ | *97* | 90 | 10 | |
| $CD_3Ti(OCHMe_2)_3$ | *6a* | 96 | 4 | |
| $CD_3Zr(OPr)_3$ | *61a* | 45 | 55 | |

. It is known that α-hydroxy substituents of carbonyl compounds have a profound influence on direction and extent of 1,2 asymmetric induction in Grignard addition [95]. Diastereoselectivity is generally opposite to what one would expect on the basis of Cram's open chain-model because chelation effects occur (Cram's cyclic model) [94,95]. In order to test whether this model applies to titanium reagents, we allowed benzoin *142* to react with two equivalents of methyltitanium triisopropoxide *6*. Of the two possible diastereomers *146* and *147*, practically only the former was observed, which means essentially complete erythro-selectivity [83]. Other titanium reagents show the same behavior [83]. In case of $CH_3MgI$ the ratio is 96:4 and has been interpreted on the basis of the cyclic model [100]. We explain the present results analogously. The chelated forms *144* and *145* effectively compete with an open-chain conformation; nucleophilic attack then occurs anti to the phenyl group in the chelates. Consideration of ring strain makes *145* having the ends of the chelate in the axial/equatorial positions more likely than *144*.

Since the Ti—O bond length (~1.7 Å, Sect. B.II) is short relative to Mg—O (2.1 Å), the result is a little surprising[10]. Furthermore, the ligands at titanium in *145* ought

[10] Certain titanium derivatives having an ethylene glycolate type of bridge (five membered ring) can be prepared but readily undergo ring opening and oligomerization [101]. We have observed similar effects in related systems [77].

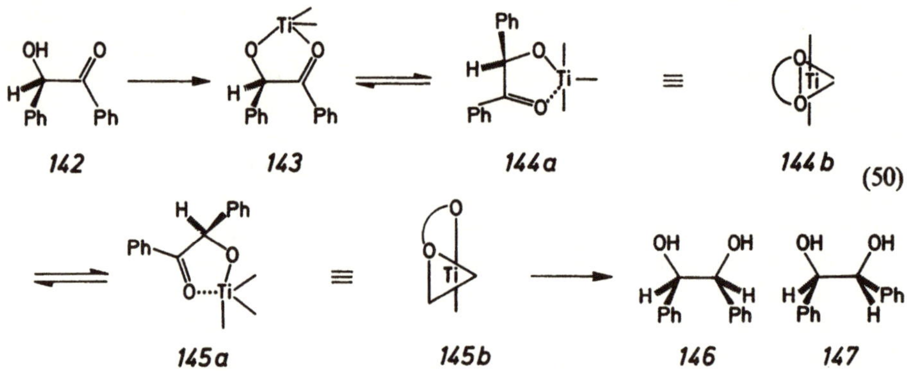


(50)

to interact sterically with the phenal group. Apparently, the latter takes on a quasi-equatorial position, minimizing 1,3 steric repulsion. If these hypotheses are correct, one would expect that an additional substituent in the five-membered chelate should cause severe 1,3 steric conjestion, since quasi-equatorial position of one group pushes the second one toward the titanium ligands. This was tested by reacting benzil *148* with two equivalents of various methylating reagents (Equation 51). Chelation control should result in meso-selectivity, as in the reactions of $CH_3Li$ and $CH_3MgCl$ (*152*:*153* = 15:85 and 28:72, respectively [100]). In complete contrast, various titanium reagents react d,l-selectively, e.g., *6* delivers a >98:2 product ratio [83]. Thus, reversal of diastereoselectivity is *not* in line with chelation *150*. Instead, an open-chain model appears to operating. We prefer to apply the Felkin-Anh model [95], i.e., *151* must be considered in which the C—OTi bond is arranged parallel to the π-system of the carbonyl group, allowing for good overlap between the $\pi^*_{CO}$ orbital and the low-lying $\sigma^*_{C-OTi}$ orbital. Antiperiplanar attack at the carbonyl group in a *non perpendicular* manner then leads to preferential formation of *152*. The reaction of methylzirconium tri-n-propoxide *61* allows for additional insight. It is meso-selective, just like the magnesium and lithium reagents *152*:*153* = 19:81) [83]! Again, it is useful to consider bond lengths. The Zr—O bond is fairly long (2.1 Å; Sect. B.II), and this makes fivemembered chelation more probable [83].

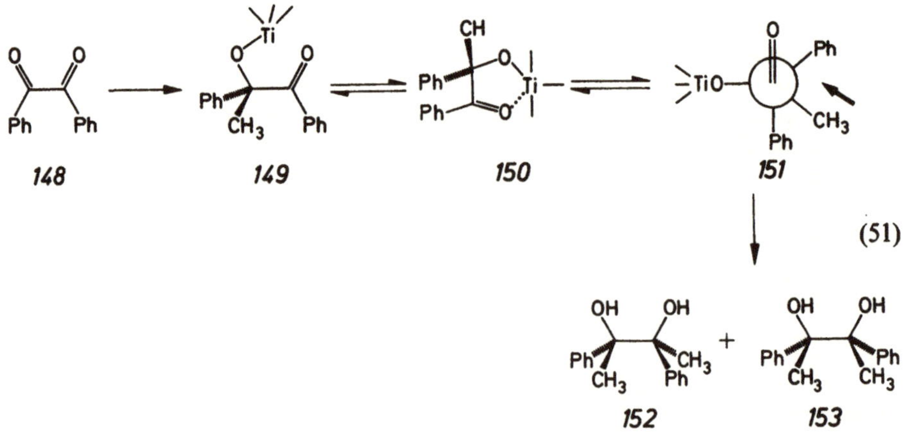


(51)

Although simple ketones or aldehydes bearing α-alkoxy groups have not yet been tested, the result of the reaction of the protected furanose *154* with methyltitanium triisopropoxide *6* is available. Essentially a single diastereomer *158* is formed (*157*:*158* = <2:98) [102]. This means reversal of diastereoselectivity relative to the analogous reaction with $CH_3MgI$ (*157*:*158* = 88:12) as reported by Hanessian [103a] and Inch [103b], who postulated complexation of magnesium with the furan and carbonyl oxygen atoms as shown in *155*. Apparently, titanium prefers *not* to undergo analogous chelation. Instead, the results are best explained by applying the Anh model as indicated in *156*. It should be noted that the working hypotheses discussed so far are based on trends currently known, and that final "rules of behavior" of titanium reagents in 1,2 asymmetric induction will be proposed as more information becomes available.

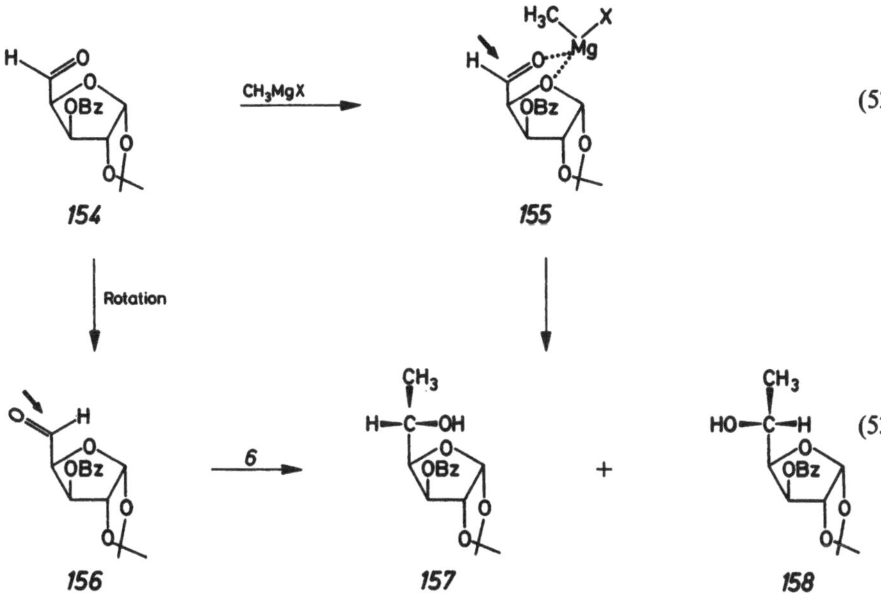


(52)

(53)

| | *157* | *158* |
|---|---|---|
| $CH_3MgX$ | 88 | 12 |
| $CH_3Ti(OCHMe_2)_3$ | <2 | >98 |

Present knowledge relating to 1,3 and 1,4 asymmetric induction involving organotitanium reagents is scanty. The reaction of the sensitive β-hydroxybutanal *159* with *6* and *14* affords 55:45 and 70:30 mixtures of meso- and d,l 2,4-pentane diols *160* and *161*, respectively [83]. Methyllithium also affords a 70:30 mixture [77] (Equation 54). Conversion to products is high in all cases (>80%).

The above level of diastereoselectivity is noteworthy because the protected form (alkylether) of *159* fails to show 1,3 asymmetric induction in the reaction with organometallics such as cuprates [104a]. As in other cases of 1,3 asymmetric induction [75a], the results are not easy to rationalize. It is likely that the reactions are initiated by deprotonation of the hydroxyl group. An *intramolecular* transfer of the

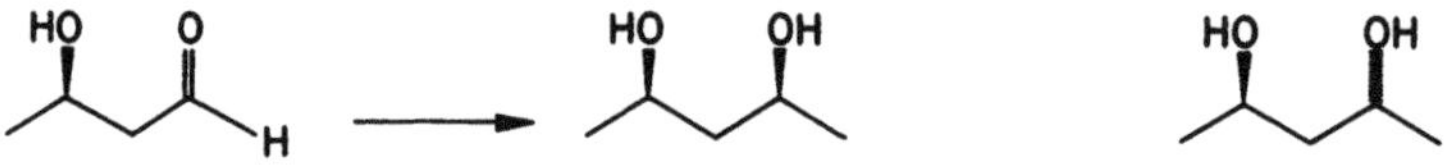


159 160 161

| | | |
|---|---|---|
| 2 $CH_3Ti(OCHMe_2)_3$ *6* | 55 : 45 | (54) |
| 1 $(CH_3)_4Ti$ *14* | 70 : 30 | |
| 2 $CH_3Li$ | 65 : 35 | |

methyl group onto the aldehyde function in case of tetramethyltitanium *14* via *162* is in line with the stereochemistry, but the bicyclo [1.1.3] transition state would be highly strained[11]. Also, it was observed that the same reaction using sixfold dilution does not affect direction or extent of diastereoselectivity [83)]. In contrast, chelation according to *163* is in line with the data[12].

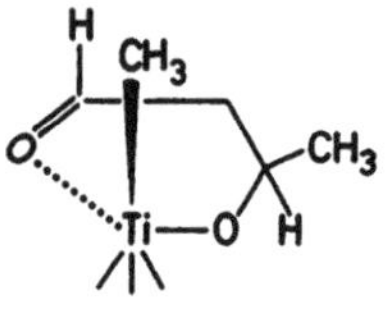


*162*

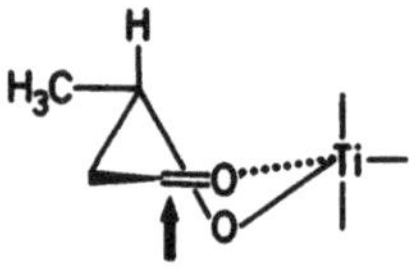


*163*

Finally, the first case of 1,4 asymmetric induction in the addition of organotitanium reagents to carbonyl compounds has been reported [83)]. O-Phthalic-dicarboxaldehyde *164* reacts with one equivalent of methyltitanium triisopropoxide *6* to form the mono-adduct *165* in high yield (>90% conversion; 71% isolated)[13]. If two equivalents of *6* are employed, smooth dimethylation occurs, providing an 83:17 mixture of d, l and meso diols *166* and *167*, respectively (Equation 56) [83)].

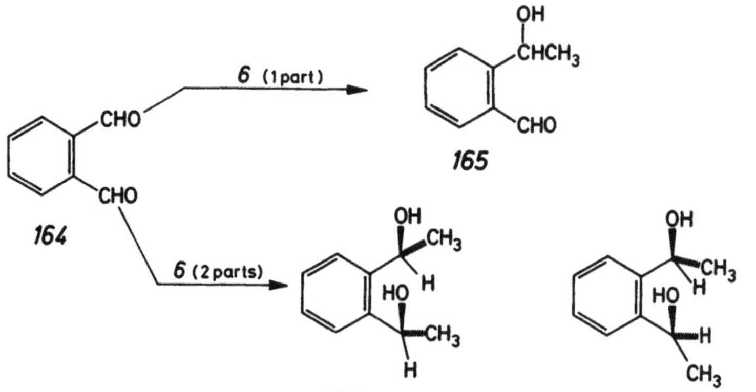


(55)

(56)

*166* *167*

[11] Allyl groups are more likely to be suited (bicyclo-[1,3,3] transition state), and this is being studied.

[12] A boat form can also be invoked [83)]. It should be noted that the direction of attack is opposite to that proposed for chelation controlled reduction of β-hydroxy ketones [104b)].

[13] *165* exists predominantly in the cyclic hemi-acetal form (not shown). Controlled mono-addition using $CH_3MgI$ is not possible, since a mixture of product results containing <50% of *165* [77)].

The analogous reaction using tetramethyltitanium *14* results in a 42:58 product ratio, whereas $CH_3MgI$ is non-selective (1:1 mixture) [83]. It is possible that in case of *14* the transfer of the second methyl group occurs to some extent intramolecularly via a bicyclo[1,1,4] transition state. Indeed, sixfold dilution (which should reduce the amount of intermolecularity) reverses diastereoselectivity, *166*:*167* being 52:48 [83]. Clearly, systematic studies are required before final conclusions regarding the use of titanium or zirconium mediated intramolecular addition reactions can be reached.

## D.II Diastereoselectivity in Addition Reactions Involving Cyclic Ketones

The problem of axial or equatorial addition of organolithium, magnesium, aluminium, zinc and cadmium reagents to cyclic ketones has been studied intensely [105]. According to Ashby and other authors, steric and torsional strain are the two most important factors which influence the direction of attack [105]. In systems in which complexation of the carbonyl oxygen occurs followed by addition of a second molecule of organometallic reagent (e.g., $(CH_3)_3Al$), the so called compression effect also operates [105]. In order to see how titanium compounds compare with the above classical reagents, systematic studies were initiated and are still in progress.

The first system to be studied was 2-methyl-2-tolyl-cyclopentanone *168*. In early 1979 we attempted to perform direct geminal dimethylation of this compound using dimethyltitanium dichloride *171*[14] in hope of obtaining curparene (Sect. F.II). Initial attempts were unsuccessful, since the product turned out to be the Grignard-type adducts *169* and *170* (Equation 57).

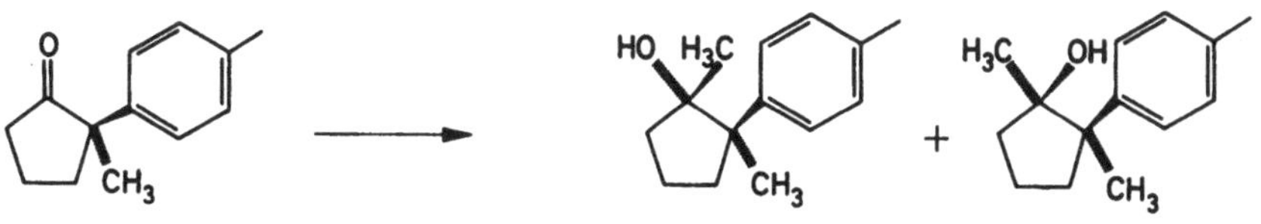


*168* *169* *170* (57)

| | *169* | *170* |
|---|---|---|
| $(CH_3)_2TiCl_2$ *171* | 95 | 5 |
| $CH_3Li$ | 34 | 66 |
| $(CH_3)_4Ti$ *14* | 89 | 11 |

What struck us at that time was the fact that the diastereomer ratio *169*:*170* was 95:5, whereas the reaction of $CH_3Li$ led to a 34:66 product distribution[15]. Indeed, this was one of the early observations which triggered our interest in using organotitanium reagents for stereo- and chemoselective C—C bond formation in a general way [71]. The $CH_3Li$ reaction is in line with preferred attack at the less hindered

[14] Prepared by reacting $(CH_3)_2Zn$ with $TiCl_4$ in $CH_2Cl_2$ (>90% yield) [41].

[15] Essentially the same results were obtained upon replacing the tolyl group by phenyl.

face of the carbonyl group[16]. Reversal of diastereoselectivity upon using *171* can be explained by postulating Lewis acid-Lewis base interaction between the π-electrons of the aryl group and the titanium reagent [22]. Such complexation would then result in preferred methylation cis to the more bulky tolyl group. This hypothesis is reminiscent of the metal-phenyl interaction in tetrabenzyltitanium (Sect. B.II). It should be noted that the ketone *168* is sterically shielded, so that the rate of addition of *171* is considerable lower than those involving other non-hindered ketones. Besides methyltitanium halides, tetramethyltitanium *14* also has a very high tendency to form stable complexes with Lewis bases (Sect. B.I). Thus, we speculated that *14* would behave more like *171*, and not like $CH_3Li$. This turned out to be the case (Equation 57) [66].

Unfortunately, the above directive effect does not dominate in all cases. 2-Phenylcyclopentanone *88* reacts with *171* to form a 20:80 product distribution of *89a* and *89b*, whereas $CH_3Li$ affords a 10:90 ratio [21]. The results obtained by using other organotitanium reagents (Equation 58) do show a certain trend, however [21,66]. The more Lewis acidic reagents tend to yield more of *89a*, indicating a competition between steric and complexation effects. In contrast to *168*, *88* reacts very rapidly. It should be noted that the total yield in case of $CH_3Li$ is <55% (Sect. C.V) due to competing deprotonation. Conversion using titanium reagents is >85%.

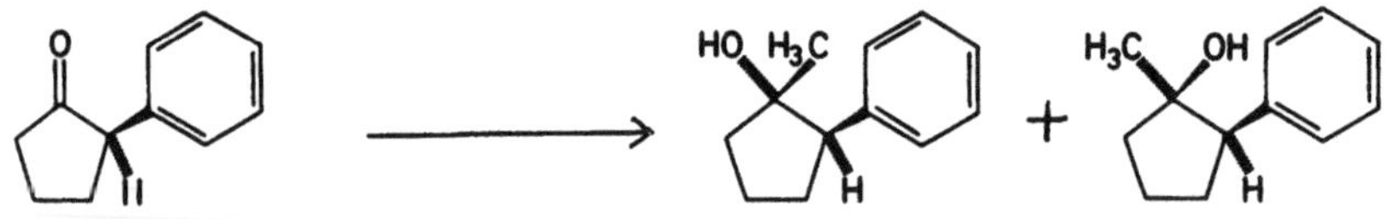

*88* *89a* *89b*

| | *89a* | *89b* |
|---|---|---|
| $CH_3Li$ | 10 | 90 |
| $(CH_3)_2TiCl_2$ *171* | 20 | 80 |
| $(CH_3)_4Ti$ *14* | 53 | 47 |
| $(CH_3)_3TiOCHMe_2$[17] *172* | 62 | 38 |
| $(CH_3)_2Ti(OCHMe_2)_2$ *63* | 3 | 97 |

(58)

Turning to six membered rings, methylation of conformationally locked 4-tert-butylcyclohexanone *173* was investigated systematically (Equation 59; Table 7). With the exception of tetramethyltitanium *14* (entry 11) and the ate complex *123* (entry 12), all of the titanium reagents undergo preferential equatorial attack. The highest degree of diastereoselectivity was observed for methyltitanium triisopropoxide in hexane (*174*:*177* = 94:6; entry 9) which compares rather well with Ashby's method [106], according to which $CH_3Li$ (two parts) is reacted with *173* (one part) in presence of $LiClO_4$ (one part) (*174*:*175* = 92:8, entry 15), or with the use of a

---

[16] This assumes that phenyl is more bulky than methyl and neglects the precise conformation of the molecule (which is not known).

[17] Prepared by reacting $Cl_3TiOCHMe_2$ with $CH_3Li$ in ether. It is not known whether this compound undergoes adduct formation with the solvent, nor has the aggregation state been determined.

mixture of $CH_3Li$ and $(CH_3)_2CuLi$ (entry 16)[107]. Interestingly, the zirconium analog *92* is less selective (entry 21)[80].

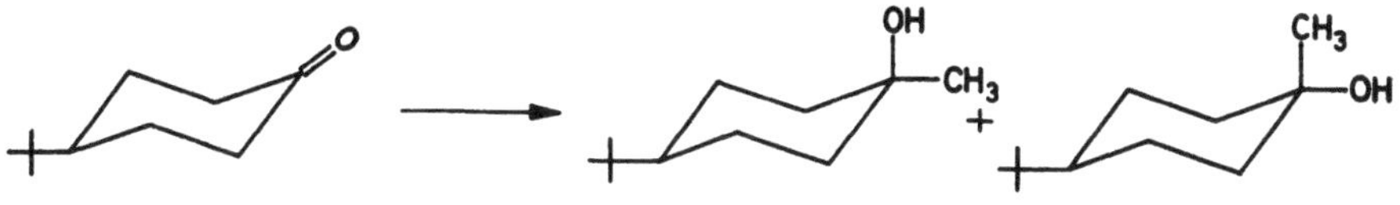

*173* *174* *175* (59)

We believe that steric interaction of the entering, bulky titanium reagent with the axial hydrogen atoms outweighs torsional effects induced by the 2,6-hydrogen atoms, as has been assumed for classical methylating agents[105]. Table 7 also reveals that solvent or possible salt effects are small. A striking difference was observed upon using tetramethyltitanium *14* (entry 11) or the are complex *123* (entry 12), the degree of attack at the sterically more hindered site being significant. This is reminiscent of Ashby's observation that a threefold excess of $(CH_3)_3Al$ also leads to reversal of diastereoselectivity (entry 18). He introduced the concept of compression, i.e., "the effective bulk of the carbonyl groups increases to such an extent

**Table 7.** Methylation of 4-tert-Butylcyclohexanone *173*

| Entry | Reagent | Solvent | Temp. (°C) | *174*:*175* | Ref. |
|---|---|---|---|---|---|
| 1. | $(CH_3)_2TiCl_2$ [a] *171* | $CH_2Cl_2$ | −78 | 82:*18* | 72) |
| 2. | $(CH_3)_2TiCl_2$ [a] 121 | $CH_2Cl_2$ | −50 | 76:24 | 35) |
| 3. | $(CH_3)_2Ti(OCHMe_2)_2$ [a] *63* | $CH_2Cl_2$ | −40 | 75:25 | 35) |
| 4. | $CH_3Ti(OCHMe_2)_3$ [b] *6* | $CH_2Cl_2$ | 22 | 82:18 | 35) |
| 5. | $CH_3Ti(OCHMe_2)_3$ [b] *6* | ether | 0 | 89:11 | 22) |
| 6. | $CH_3Ti(OCHMe_2)_3$ [b] *6* | ether | 22 | 86:14 | 35) |
| 7. | $CH_3Ti(OCHMe_2)_3$ *6* | ether | ? | 86:14 | 76) |
| 8. | $CH_3Ti(OCHMe_2)_3$ *6* | $CH_3CN$ | ? | 86:14 | 76) |
| 9. | $CH_3Ti(OCHMe_2)_3$ [b] *6* | hexane | −15→22 (24 h) | 94: 6 | 35) |
| 10. | $CH_3Ti(OCHMe_2)_3$ [c] *6* + LiCl | ether | 22 | 93: 7 | 35) |
| 11. | $(CH_4)_4Ti$ [a,c] *14* + LiCl | ether | −50 | 38:62 | 35) |
| 12. | $CH_3\overset{\ominus}{Ti}(OCHMe_2)_4\overset{\oplus}{Mg}I$ *123* | ether | −40 | 33:67 | 77) |
| 13. | $CH_3MgI$ | ether | 0–5 | 62:38 | 103) |
| 14. | $CH_3Li$ | ether | 5 | 65:35 | 103) |
| 15. | 2 $CH_3Li$ + $LiClO_4$ | ether | −78 | 92: 8 | 105) |
| 16. | 2 $CH_3Li$ + 3 $(CH_3)_2CuLi$ | ether | −70 | 94: 6 | 104) |
| 17. | $(CH_3)_3Al$(1 part) | benzene | 22 | 76:24 | 103) |
| 18. | $(CH_3)_3Al$(3 parts) | benzene | 22 | 12:88 | 103) |
| 19. | $CH_3CdCl$ + $MgX_2$ | ether | 0–5 | 38:62 | 103) |
| 20. | $(CH_3)_2Zn$ + $MgX_2$ | ether | 0–5 | 38:62 | 103) |
| 21. | $CH_3Zr(OBu)_3$ [c] + LiCl | ether | 22 | 80:20 | 80) |

[a] Amount of reagent chosen so that only one methyl group was transfered
[b] The reagent in entries 4, 5, 6, (presumably 7, 8) and 9 was distilled, i.e., no salts were present during addition to *173*
[c] In situ mode of reaction, i.e., LiCl was present

by complexation with an organoaluminium compound that severe interaction with groups on adjacent carbons can occur in the transition state" [105]. Phenomena similar to the compression effect may be operating to some extent in case of *14* and *123*[18].

Allylmetal reagents are known to add to *173* preferentially from the axial direction (equatorial alcohol) because torsional strain outweighs 1,3 axial steric strain [105]. For example, allylmagnesium bromide affords a 55:45 mixture of axial and equatorial alcohol, respectively [105], Allyltitanium tris(diethylamide) *102* is more selective (80:20 mixture of equatorial and axial alcohol) [90]. Phenyltitanium triisopropoxide *5* affords a 1:1 mixture of diastereomers [78], much like the lithium and magnesium analogs [105]. Thus, this is another case in which titanation is of no help in increasing diastereoselectivity (see also Equation 48).

Organometallics attack 2-methylcyclohexanone *176* preferentially from the equatorial direction because the pseudo-axial hydrogen of the methyl group contributes to 1,3 steric repulsion [105]. Methylation using two parts of $CH_3Li$ and one part $LiClO_4$ per one part *176* is highly selective [106] (Equation 60). *6* (in situ from $CH_3Li$ and *3*) affords a 94:6 mixture [83].

Similar results pertain to phenyllithium [106] and the phenyltitanium reagent *5* (Equation 60) [66]. If the side chain in cyclohexanone is sterically more demanding, titanium reagents are completely diastereoselective (see Equation 17).

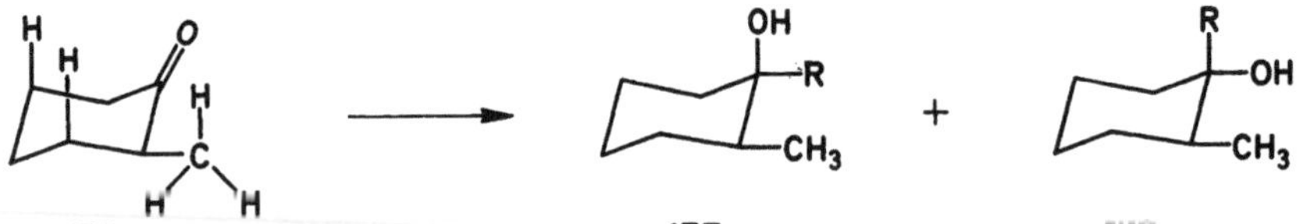


*176* *177* *178*

| | 177 | 178 | |
|---|---|---|---|
| $CH_3Ti(OCHMe_2)_3$ *6* | 94 | 6 | |
| $2CH_3Li + LiClO_4$[106] | 94 | 6 | |
| $PhTi(OCHMe_2)_3$ *5* | >95 | <5 | (60) |
| $2PhLi + LiClO_4$[106] | 96 | 4 | |
| $(CH_2{=}CHCH_2)_2Mg$[105] | 76 | 24 | |
| $Cp_2Ti$–allyl[19] | 81 | 19 | |
| $CH_2{=}CHCH_2Ti(OCHMe_2)_3$ *96* | 76 | 24 | |

Conformationally mobile 3-methyl- and 4-methylcyclohexanone *179* and *182*, respectively, react with $CH_3MgI$ (2 h, 22 °C) essentially non-selective, in contrast to *6* (distilled; reactions in hexane, −15 °C→22 °C, 24 h) [77] as shown in Equations *61* and *62*.

[18] This aspect has not yet been looked at systematically, e.g., by using an excess of *14* or *123*.

[19] This is a Ti(III)-reagent prepared from $Cp_2TiCl_2$ and two equivalents of allylmagnesium chloride [22].

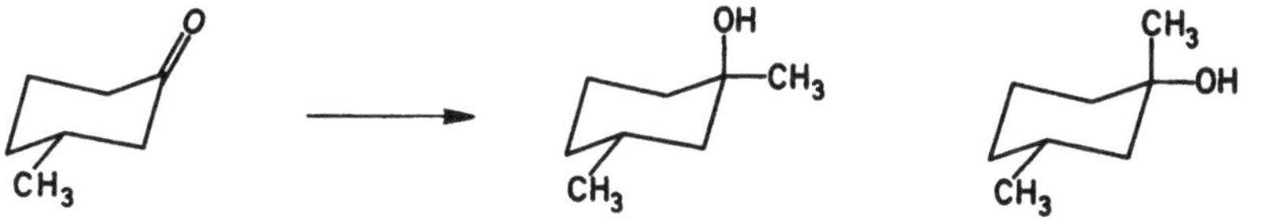

179 180 181 (61)

| | 180 | 181 |
|---|---|---|
| $CH_3MgI$ | 54 | 46 |
| $CH_3Ti(OCHMe_2)_3$ *6* | 89 | 11 |

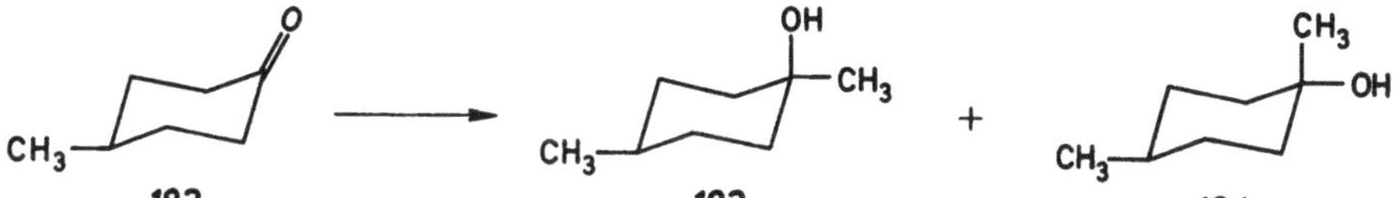

182 183 184

| | 183 | 184 |
|---|---|---|
| $CH_3MgI$ | 52 | 48 |
| $CH_3Ti(OCHMe_2)_3$ *6* | 88 | 12 |

(62)

## D.III Diastereoselective Aldol-Type Additions

Nucleophiles such as enolates or substituted allylmetal compounds are known to react with prochiral aldehydes and ketones to form mixtures of threo or erythro adducts. In case of aldehydes, high degrees of diastereoselection have been achieved [48,91,108)]. In the following three Sections, reactions of titanium and zirconium enolates as well as allyl derivatives are presented.

### D.III.1 Titanium Enolates in Aldol Additions

The problem of diastereoselective aldol addition has been largely solved [48,108)]. Under kinetic control Z enolates favor erythro adducts and E enolates the threo diastereomers, although exceptions are known. This has been explained on the basis of a six-membered chair transition state in which the faces of the reaction partners are oriented so as to minimize 1,3 axial steric interactions [48,108)]. This means that there is no simple way to prepare erythro aldols from cyclic ketones, since the enolates are geometrically fixed in the E geometry.

We have prepared a number of titanium enolates by quenching the lithium analogs with chlorotitanium triisopropoxide *3*, chloro- or bromotitanium tris(diethylamide) *15* or chlorotitanium tris(dimethylamide) *103* [25)]. In most cases the solutions can be freed from the ether or THF and the stable liquid titanium enolates studied by NMR if so desired. In case of the amino derivative *186* flash distillation is possible. The H-NMR spectra of the distilled and non-distilled compound are identical and prove the existence of genuine titanium enolates[20]. For synthetic purposes an in situ reaction mode is the method of choice.

*185* → *186* (63)

A number of other acyclic Z and E lithium enolates were quenched similarly. In all cases the stereochemistry at the enol double bond was retained, as shown by subsequent conversion into the corresponding silyl enol ether. Upon reacting the titanium enolates with aldehydes, very clean aldol addition occured (>90% conversion at −78 °C). Generally, erythro-selectivity was observed *irrespective of the geometry of the enolate*. Equations *64* and *65* are typical [25].

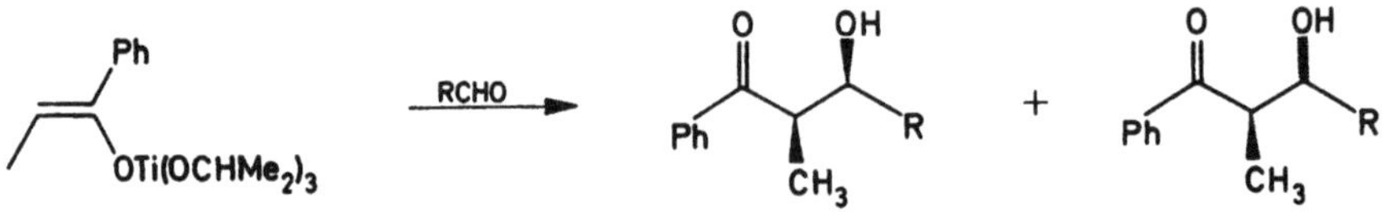

| *187* | *188* | *189* |
|---|---|---|
| R = Ph | 87 | 13 |
| R = Et | 89 | 11 |

(64)

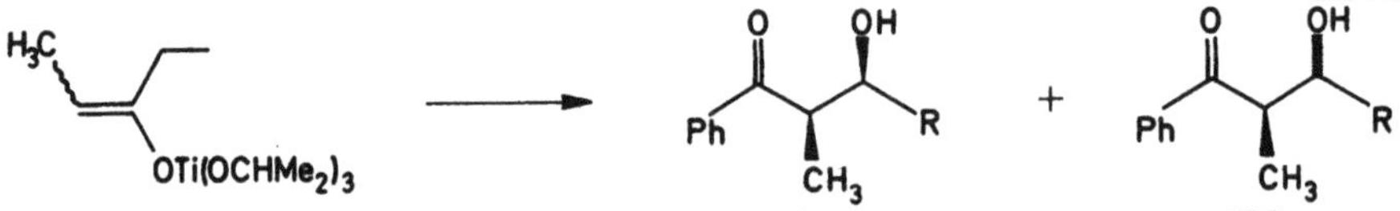

| *190* | *191* | *192* |
|---|---|---|
| Z : E = 36 : 64 | 89 | 11 |
| Z : E = 92 : 8 | 88 | 12 |

(65)

In isolated cases diastereoselectivity is reversed [25]. Either acyclic transition states are transversed, or the usual pericycles are involved, having chair or boat conformations depending upon the enolate geometry, the substituents and the nature of the aldehyde [25]. These possibilities have also been discussed in "irregularities" reported for other enolates [48,109–111]. For example, dicyclopentadienylchlorozirconium- [109], triphenyltin [110] and tris(dialkylamino)-sulfonium [111] (TAS) enolates also favor

[20] *186* contains less than 1% of the E isomer. H-NMR($CDCl_3$): $\delta = 0.97$ (t, 18H), 1.67 (d, 3H), 3.45 (q, 12H), 4.83 (q, 1H), 7.0–7.53 (m, 5H); $^{13}$C-NMR ($CDCl_3$): $\delta = 11.4$, 15.1, 46.6, 98.1, 125.2, 126.6, 127.6, 140.8, 159.1. In lit [25] it was mistakenly stated that the triisopropxy analog of *186* is distillable. Generally, only tris-amino derivatives are thermally stable enough for such isolation procedures.

erythro product formation irrespective of the enolate geometry as reported by Evans, Yamamoto and Noyori. Since the diastereoselectivities in these reactions are seldom above 90%, the advantage of not having to prepare prochirally pure enolates is not that apparent. The situation in cyclic ketones is a little different, since the corresponding lithium, magnesium and boron E enolates react threo-selectively. Ti enolates react erythro-selectively, thereby filling a synthetic gap, e.g., Equation 66 [25].

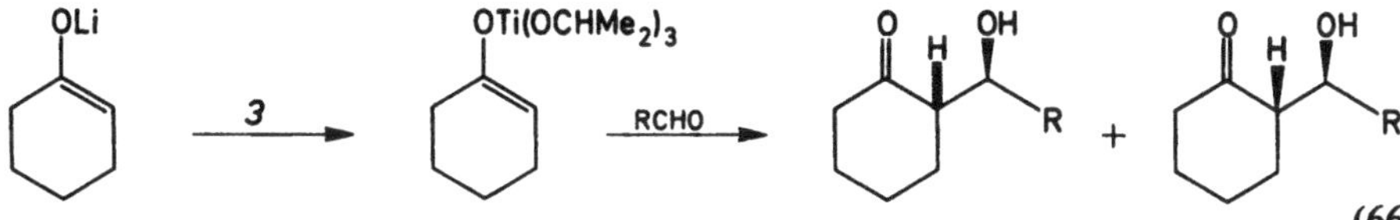


*193* *194* *195* *196* (66)

| | 195 | 196 |
|---|---|---|
| *a* R = $C_6H_5$ | 86 | 14 |
| *b* R = $CH(CH_3)_2$ | 96 | 4 |

Variation of the ligands at titanium can lead to improvements. For example, enolates *197a–b* react with benzaldehyde to afford 92:8 and 97:3 mixtures of *195a* and *196a*, respectively [25]. Enolates *199–202*, derived from ketones, lactones and lactams, add erythro-selectively to benzaldehyde to yield the corresponding aldols (erythro:threo as 85:15, 91:9, 92:8, 58:42 and 88:12, respectively) [25,77].

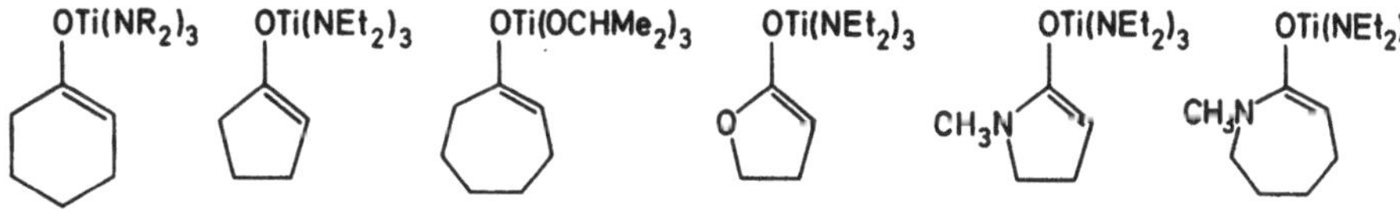


*197* *198* *199* *200* *201* *202*

*a* R = Me
*b* R = Et

On the basis of the present data, a preliminary conclusion regarding the mechanism of the above reactions can be reached. Since the diastereoselectivity is significantly increased in going from *197a* to *197b*, an acyclic transition state seems unlikely. We therefore prefer to postulate a boat transition state [25]. The effect of further ligand variation remains to be studied. Initial experiments in which isopropoxy ligands are replaced by phenoxy show clear improvements, selectivities being 90–98% [77].

Trialkoxyzirconium enolates obtained from *193* show similar degrees of erythroselectivity [25,77]. Dicyclopentadienylchlorozirconium enolates are less selective [109]. However, in what appears to be a highly interesting approach, Evans has reported that dicyclopentadienylchlorozirconium enolates derived from prolinol amides exhibit not only excellent diastereoselection (erythro preference), but also enantioselectivity [112]. Related titanium enolates having *one* cyclopentadienyl ligand show lower degrees of erythroselectivity [77]. Cyclic TAS enolates have been studied only in one case, i.e., in the reaction of cyclopentanone enolate with isobutyraldehyde which proceeds completely erythroselectively if short reaction times are used [111].

Not much is currently known concerning diastereoselective addition of metal enolates to *ketones* [48, 108], but selectivities are expected to be lower. In case of titanium enolates, several examples have been studied [77]. The reaction shown in Equation 67 involves an ester-enolate[21] and proceeds strictly in a 1,2 manner with ~90% diastereoselectivity. The observation is significant because similar reactions with aldehydes are essentially stereo-random [77]. Also, the lithium analog of *203* affords a 1:1 mixture of diastereomers. Diastereoface-selectivity in Equation 67 is not an exception, because *203* adds to acetophenone and pinacolone to afford 85:15 and >76:24 diastereomer mixtures, respectively [77]. Although stereochemical assignments have not been made in all cases, the acetophenone adduct was converted stereospecifically into the β-lactone which was decarboxylated to yield an 85:15 mixture of Z- and E-2-phenyl-2-butene [77].

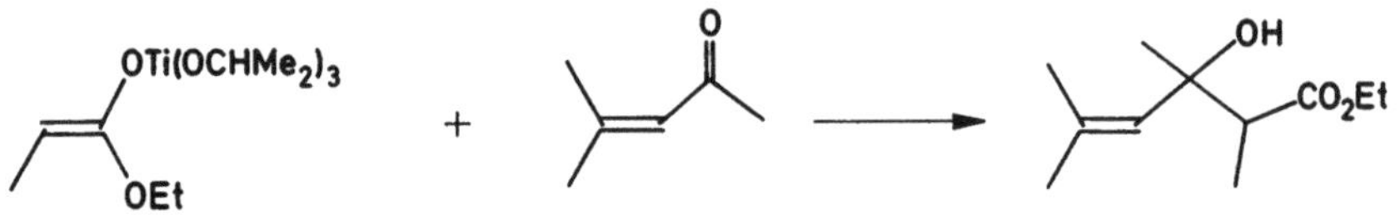

(67)

*203* *204* *205*

Michael additions of organotitanium or zirconium reagents remain to be explored systematically. Recently, Stork described an interesting stereoselective intramolecular Michael addition in which zirconium enolates appear to be involved [113]. In another Michael type process, methyltitanium triisopropoxide *6* was added enantioselectively to a chiral α,β-unsaturated sulfoxide, but $CH_3MgCl$ was more efficient [114].

## D.III.2 Aldol-Type Addition of Titanated Hydrazones

Metal enolates derived from *aldehydes* have not been used successfully in diastereoface-selective aldol additions [48, 108]. Stereoselection is expected to be lower than in case of ketone or ester enolates, because steric interactions in pericyclic transition states are not as pronounced. Thus, ketone or ester enolates have to be used, and the adducts chemically modified to provide the desired aldols [48, 108]. Whereas titanium enolates prepared from aldehydes are also not very diastereoselective [77], the corresponding titanated hydrazones display pronounced erythro-selectivity [115]. They are prepared by quenching lithiated hydrazones [116] with *3* or *15* and turn out to be isolable air sensitive compounds, e.g., Equation 68. In case of *208* the H-NMR spectrum was recorded which shows that titanium is bonded to nitrogen and that the double bond has E configuration ($J_{CH-CH} = 16.9$ Hz). The lithium precursor *207* also has E configuration as well as Z configuration about the C—N bond. This latter type a stereochemistry does not apply to *208*, since the Ti—N bond is covalent and inversion barriers are low [115].

---

[21] *203* was prepared by quenching the lithium enolate (95:5 trans/cis mixture) with *3* in THF. Aldol addition was performed at −78 °C (0.5 h), conversion to *205* being >90% (80% isolated by distillation).

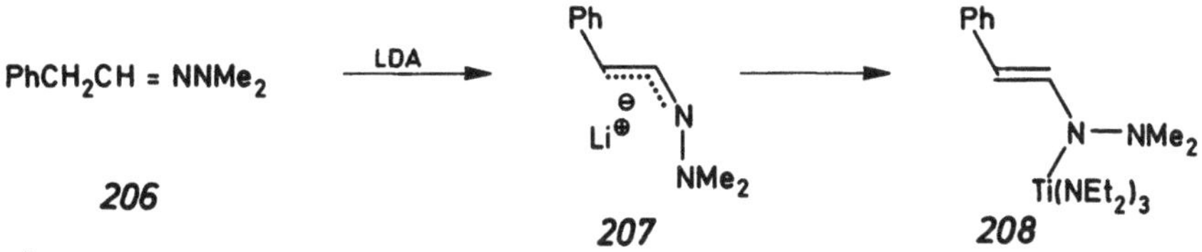


(68)

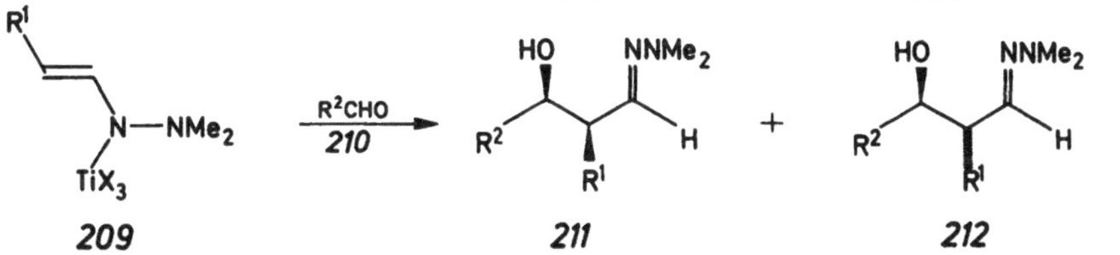


(69)

Reactions of titanated hydrazones with aldehydes occur cleanly at —20 °C (Equation 69). It is not clear whether the observed erythro-selectivity (Table 8) depends upon the geometry of the double bond, since attempts to prepare Z configurated analogs were not rewarding [115]. The lithiated precursors themselves are unsuitable for selective additions. Titanation not only increases stereo-differentiation, but also chemoselectivity [115]. The assumption of a pericyclic transition state means that chair orientations *213* and *214* must be considered, the latter being of higher energy. However, boat transition states also explain the results [115].

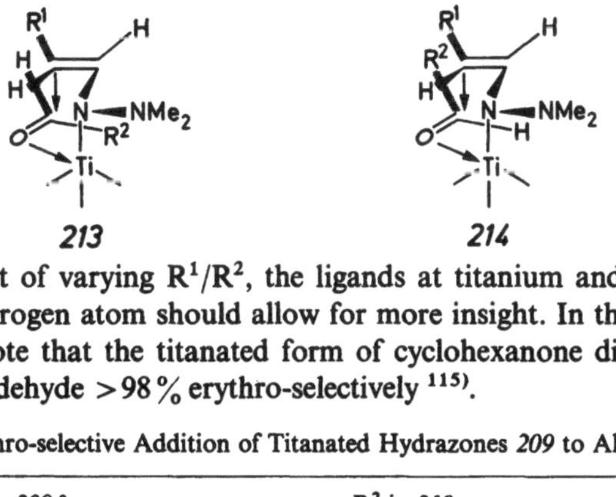


*213* *214*

The effect of varying $R^1/R^2$, the ligands at titanium and the substituents at the terminal nitrogen atom should allow for more insight. In this connection it is interesting to note that the titanated form of cyclohexanone dimethylhydrazone reacts with benzaldehyde >98% erythro-selectively [115].

**Table 8.** Erythro-selective Addition of Titanated Hydrazones *209* to Aldehydes

| Substituents in *209* [a] | | $R^2$ in *210* | *211*:*212* |
|---|---|---|---|
| $R^1$ | X | | |
| $CH_3$ | $NEt_2$ | $C_6H_5$ | 85:15 |
| $CH_3$ | $O^iPr$ | $C_6H_5$ | 91: 9 |
| $CH_3$ | $O^iPr$ | $CH_3$ | 95: 5 |
| $CH_3$ | $O^iPr$ | $C(CH_3)_3$ | 93: 7 |
| $(CH_3)_2CH$ | $O^iPr$ | $C_6H_5$ | 94: 6 |
| $C_6H_5$ | $O^iPr$ | $C_6H_5$ | 98: 2 |
| $C_6H_5$ | $NEt_2$ | $CH_3$ | 91: 9 |
| $C_6H_5$ | $O^iPr$ | $CH_3$ | 96: 4 |
| $C_6H_5$ | $O^iPr$ | p-$NO_2$—$C_6H_5$ | 98: 2 |
| $C_6H_5$ | $O^iPr$ | $(CH_3)_2CH$ | >97: 3 |

[a] In case of triisopropoxide derivatives, some degree of ligand redistribution occurs, so that more than one titanium species is present.

## D.III.3 Addition of Substituted Allyltitanium Reagents to Aldehydes and Ketones

A great deal of interest has evolved around diastereoselective formation of β-methylhomoallyl alcohols, because this structural unit is potentially useful in the synthesis of macrolide antibiotics and other compounds [48,108]. Several but-2-enyl-metal reagents have been added to aldehydes threo- or erythro-selectively (Equation 70). Diastereoselectivity depends upon whether the E or Z compounds are used (e.g., boron reagents) [91,117], but in certain cases it is independent of the geometry (erythro-selective tin reagents) [118]. Lithium or magnesium reagents do not show significant degrees of selectivity [91]. In order to test how titanium reagents perform, *219* and *220* as well as the ate complexes *114* and *221* were reacted with a number of aldehydes *216* at −78 °C [90], (Table 9). Sandwich compounds *222–224* were also employed [119–121].

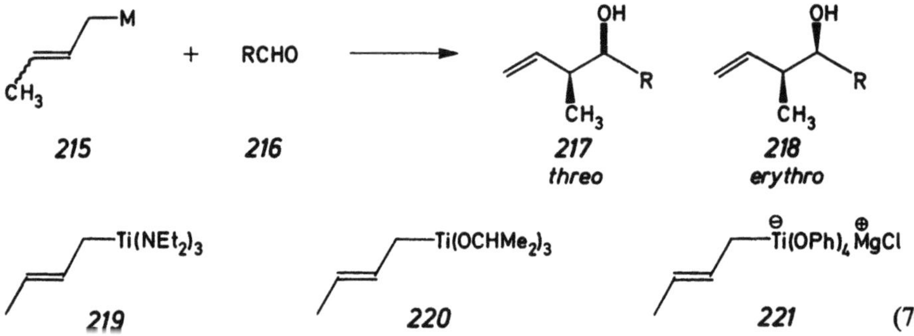


*215* *216* *217* *threo* *218* *erythro*

*219* *220* *221* (70)

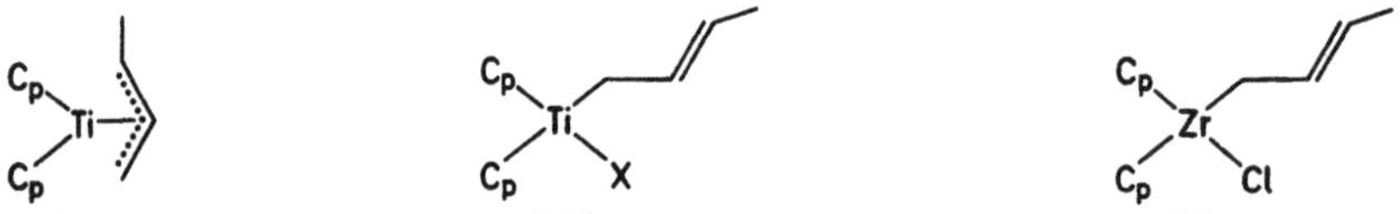


*222* *223* *224*

*a* X=Cl *b* X=Br *c* X=I

**Table 9.** Threo-selective Addition of Crotyltitanium and Zirconium Reagents to Aldehydes *216*

| Reagent | R in *216* | *217*:*218* | Ref. |
|---|---|---|---|
| *114* | $C_6H_5$ | 84:16 | 90) |
| *219* | $C_6H_5$ | 69:31 | 90) |
| *219* | n-$C_6H_{13}$ | 82:18 | 90) |
| *219* | $CH_3CH_2$ | 85:15 | 90) |
| *219* | c-$C_6H_5$ | 88:12 | 90) |
| *220* | $C_6H_5$ | 80:20 | 90) |
| *221* | $C_6H_5$ | 66:34 | 90) |
| *222* | $C_6H_5$ | 95: 5 | 118) |
| *222* | $CH_3CH_2$ | 93: 7 | 118) |
| *223a* | $C_6H_5$ | 60:40 | 119) |
| *223b* | $C_6H_5$ | 100: 0 | 119) |
| *223c* | $C_6H_5$ | 94: 6 | 119) |
| *224* | $C_6H_5$ | 81:19 | 120) |

The best results affre obtained by using Sato's sandwich compounds *223* [119]. However, the nature of the ligands in reagents of the type *219–220* have not been varied systematically.

Quenching the anion *225* with titanium tetraisopropoxide *2* affords the ate complex *226*, which reacts 100% regio- and distereoselectively with aldehydes to afford the threo adducts *227* (>95% conversion, ~80% isolated by distillation) [90] (Equation 71). This methodology is simpler than analogous reactions of boron reagents [122]. Furthermore, even *ketones* react threo-selectively (Equation 72) [90]. Since the adducts can be converted by acids (anti elimination) or KH (syn elimination) into the two possible diastereomeric dienes [122], the sequence is synthetically useful.

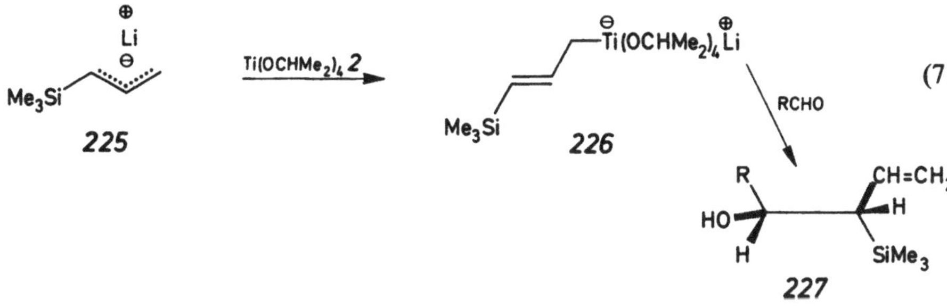


(71)

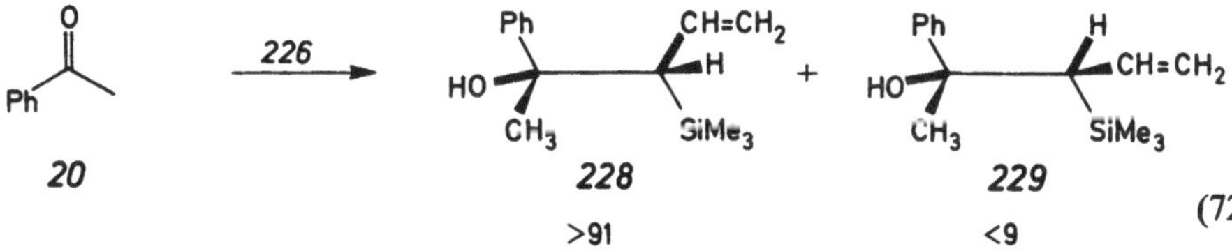


(72)

Finally, recent application of organotitanium chemistry to the problem of stereoselective homoaldol condensation (Equation 73) turned out to be highly successful, as reported by Hoppe [123]. The threo/erythro ratios (*232*:*233*) were better than 97:3 for various aldehydes.

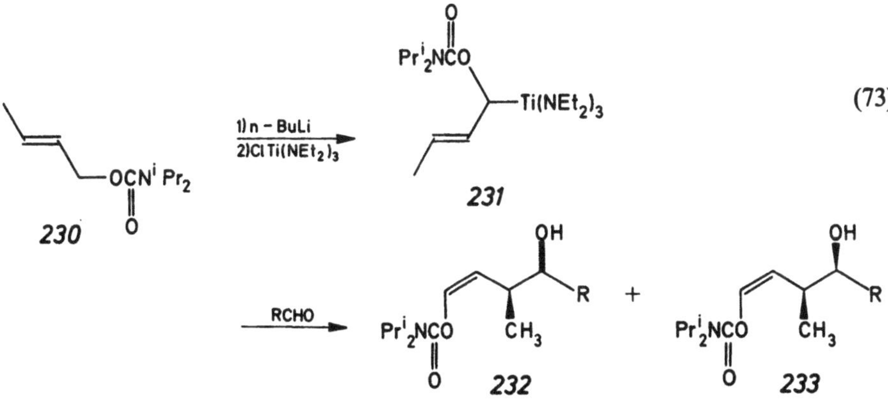


(73)

# E. Enantioselective Addition of Chirally Modified Organotitanium Reagents to Carbonyl Compounds

Application of organotitanium chemistry to enantioselective carbonyl addition is in an early stage, but promises[22] to be successful [21,78,124)]. Several approaches to the synthesis of titanium reagents having chiral ligands are possible. Addition of dialkyltitanium diisopropoxide to chiral ketones, aldehydes or alcohols yields chirally modified alkylating agents directly [21)], e.g., Equation 74. *235* reacts with benzaldehyde to form predominantly S-1-phenylethanol (ee = 13.5%) [21)].

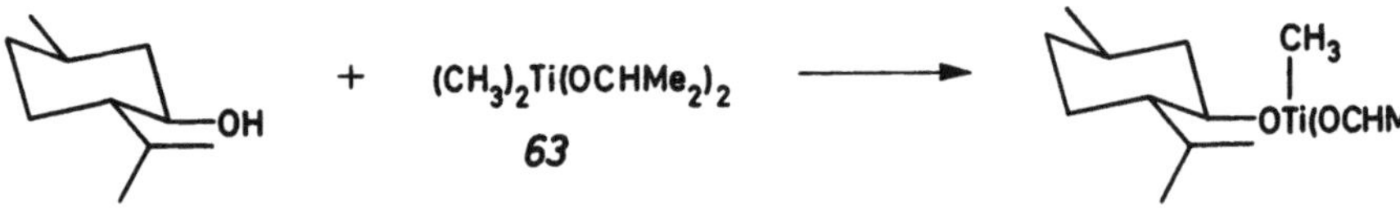


(74)

*234* *235*

Alternatively, classical organometallic reagents can be titanated by chirally modified quenching reagents as shown in Equation 75 (R* = optically active group) [35,124)]. Or course, more than one chiral alkoxy group or bidentate ligands can also be used, and chiral amino or phosphino groups are likewise potential ligands.

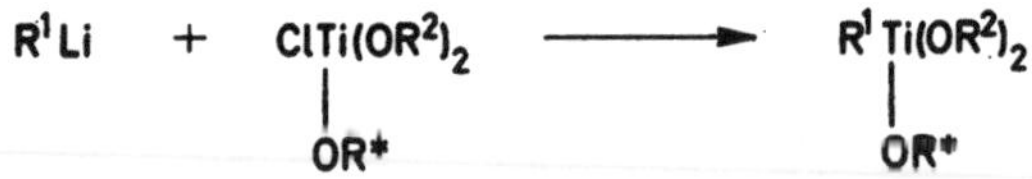


(75)

*236* *237* *238*

The currently best results pertain to the (S)-binaphthol ligand system *239* [35,124)]. Addition to benzaldehyde and aliphatic aldehydes occurs with ee-values generally ranging from 5 to 60%. Phenyl groups have also been added (ee = 20–88%) [124)]. The reactions are highly sensitive to solvent and salt effects, since not only the magnitude but also the *sign* of optical rotation changes upon varying such parameters.[23] Exploratory experiments using such compounds as *240* and *241* point to levels of enantioselectivities in the range 10–30% [35)].

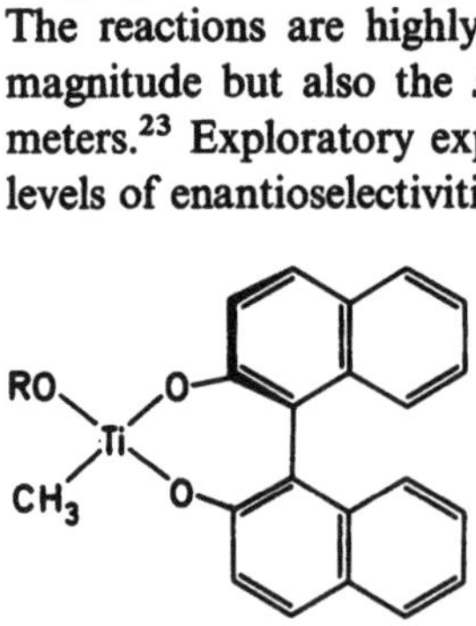


*239*

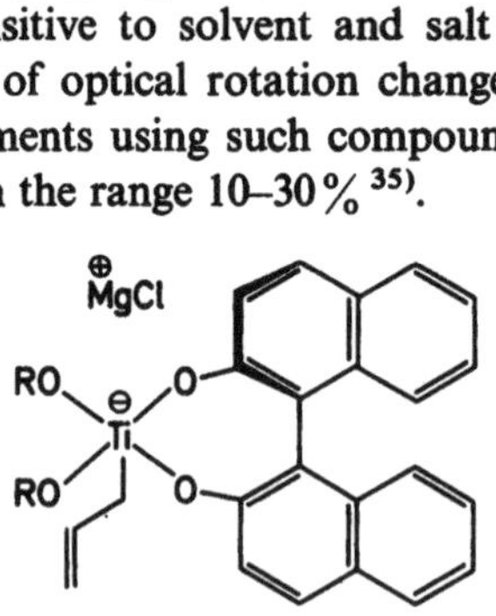


*240*

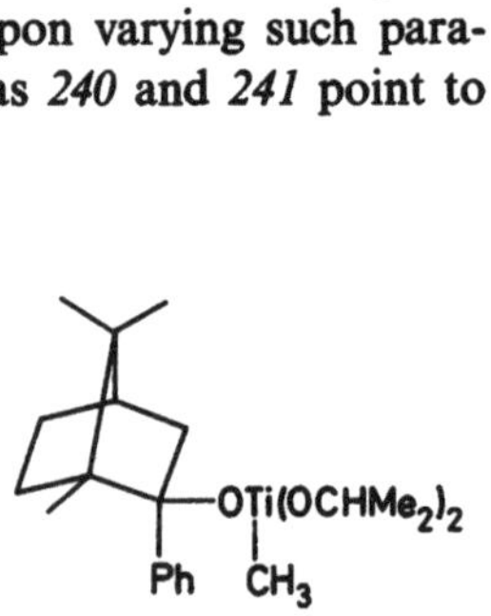


*241*

[22] For an unusual definition of "pessimism" see Ref. [124)]

[23] H-NMR studies of some of the compounds believed to be *239* [124)] are not consistent with the structure shown [35)]. Either aggregates are involved, or the seven membered ring has opened to form oligomers. These observations may explain some of the erratic results [35,124)].

A priori one could expect reagents having the chiral center at *titanium* to be particularly effective. This goal is difficult to reach because such compounds undergo racemizing ligand exchange (redistribution) reactions [35]. However, we believe that it is only a matter of time before such reagents having the right type of ligands become availalble.

# F. Reaction Types

## F.I Methylation of Tertiary Alkyl Halides

Geminal dialkylation of ketones is a process in which oxygen is replaced position specifically by two carbon residues [13,125]. If no functionality in the entering groups is desired, the sequence outlined in Equation 76 appears attractive. Whereas Grignard addition followed by formation of the tertiary alkyl chloride poses no problems, the crucial methylation cannot be performed with such basic reagents as $CH_3Li$, $(CH_3)_2CuLi$ or $CH_3MgX$ [41]. Dimethylzinc in hydrocarbon solvents has been reported to yield methylation products to the extent of 20–50% [126]. Trimethylaluminium appears to be better suited, but the synthetic scope is unknown [127]. We have discovered that methyltitanium chloride *17* or dimethyltitanium dichloride *171* are excellent methylating reagents (—78 to —20 °C, 10–30 min; $CH_2Cl_2$ as solvent), conversion generally being >90% (Equation 76) [41]. Alternatively, methylation is equally smooth if dimethylzinc and catalytic amounts of $TiCl_4$ are used. This combination generates *171* which reacts in situ with the alkyl chloride [41].

*242* *243* *244* (76)

A large number of tertiary alkyl chlorides has been methylated. Equations 77–79 are typical [41,74]. The simple synthesis of (±)-cuparene *252* is an application in terpene chemistry [35].

*245* *246* (77)

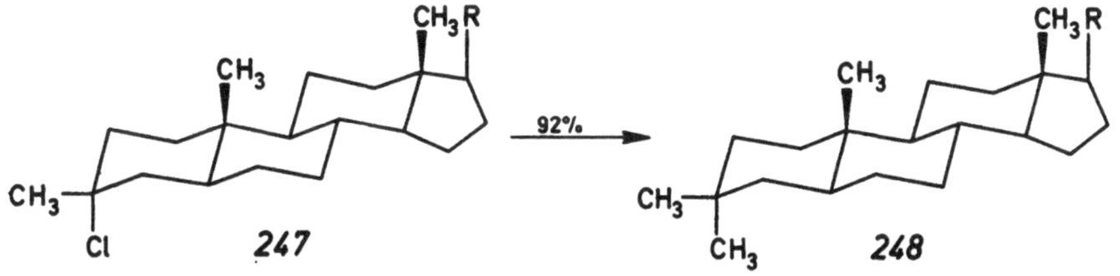


(78)

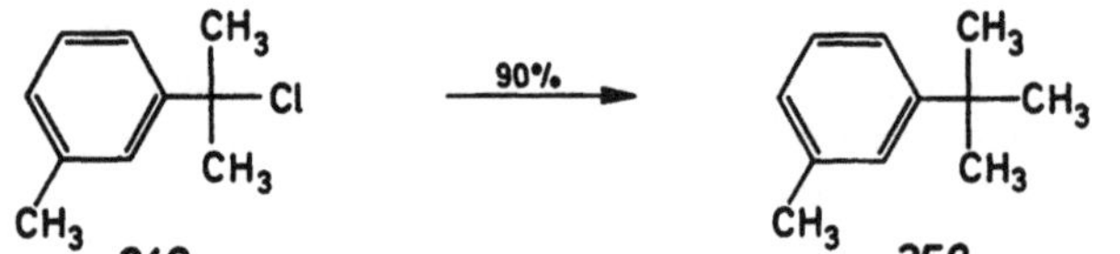


249 250 (79)

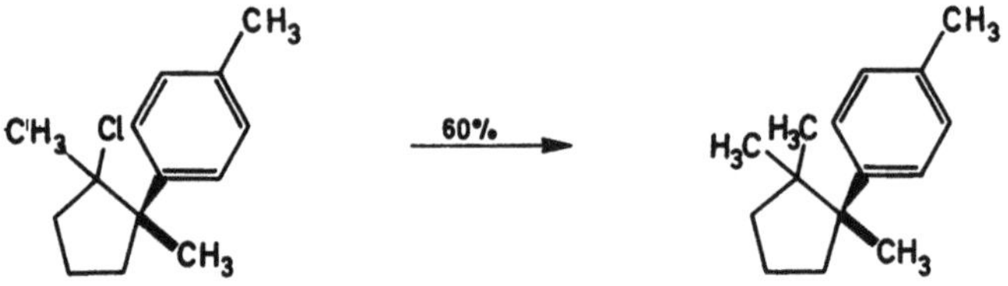


251 252 (80)

Several additional groups such as primary or secondary alkylhalide entities, double bonds, ester and methoxy groups are tolerated [35, 74], but not such basic moieties as amino groups or thioketal structural units [35]. Methylation is generally 100% position specific, although one exception has been noted; this is consistent with a carbonium ion mechanism [74]. Compounds containing two neighboring quarternary carbon atoms are also accessible [41, 74] (e.g., *252*).

Finally, the procedure allows for a new and simple way to introduce angular methyl groups (Equation 81) [128]. The fact that a mixture of diastereomers (*254*:*255* = = *60*:*40*; conversion >90%) is obtained again points to a carbonium ion mechanism. In this system we have also tested $(CH_3)_2Zn/ZnCl_2$ and $(CH_3)_3Al$ in *methylene chloride* on a preparative scale, and obtained rather similar results [35]. Here, organozinc and aluminium reagents behave much like the titanium analogs, provided one works in methylene chloride. However, striking differences are observed in related reactions as described in Section F.II. In all of the reactions a disadvantage has to do with the fact that pyrophoric reagents are required.

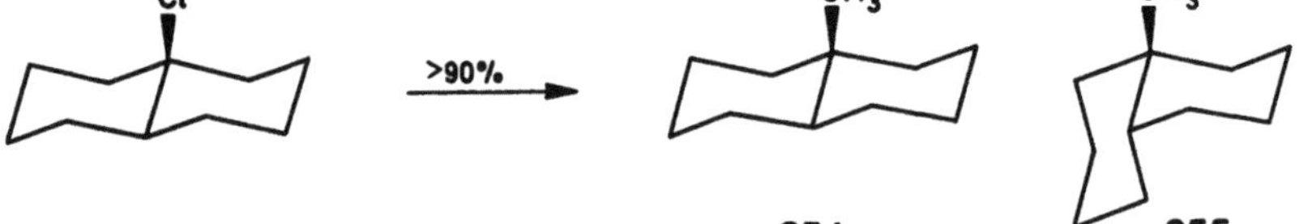


253 254 255 (81)

n-Alkyl groups cannot be introduced via titanium reagents, since β-hydride transfer onto the carbonium ions dominates [128]. In these cases dialkylzinc reagents are the reagents of choice (45–50% of isolated products in addition to reduction) [128]. In case of alkyllithium reagents, ab initio calculations relating to β-hydride abstraction by carbonium ions point to a pronounced β-effect [129], and this is also likely in titanium analogs.

## F.II Direct Geminal Dimethylation of Ketones

Dialkylation of ketones according to the above methodology is most important in case of two methyl groups because the geminal dimethyl moiety occurs frequently in terpenes, steroids and other compounds of practical and theoretical interest. Thus, efforts were undertaken to simplify the three step sequence. A novel solution to this problem involves the reaction of a ketone (one part) with *171* (two parts), yielding *256* under mild conditions (—20° to 0 °C) as shown in Equation 82 [130]. The driving force of this reaction is the formation of the strong Ti—O bond (Sect. B. III). Compounds *257–260* are typical products. Cuparene *252* has been synthesized by this method [128]. A clear limitation pertains to compounds having additional Lewis basic functional groups (amino groups, thioketal units, etc.). Nevertheless, the procedure is better than pyrolizing ketones with a large excess of $(CH_3)_3Al$ at 120° to 180 °C in a closed vessel [131]. Various combinations of $(CH_3)_3Al$ and $AlCl_3$ are also ineffective [130]. However, $(CH_3)_3Al$ and $TiCl_4$ induce direct dimethylation [130], although the yields are a little lower than in the pure titanium system. Previous methodologies for geminal dimethylation are based on multistep procedures [132], but may be preferred in certain cases in which the substrate contains a great deal of additional functionality. However, direct dimethylation according to Equation 82 in other situations should be more efficient than multistep methods, e.g., in the synthesis of certain tert-butyl aromatics [132e].

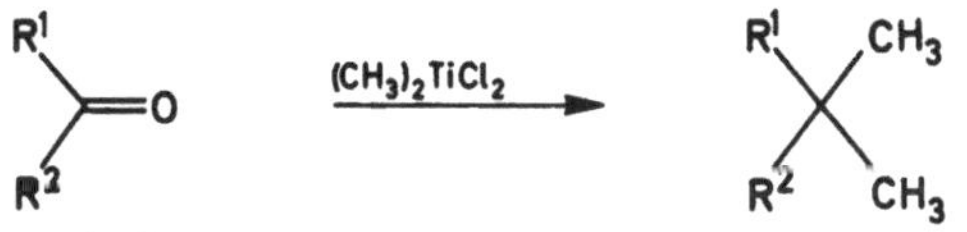


242 256 (82)

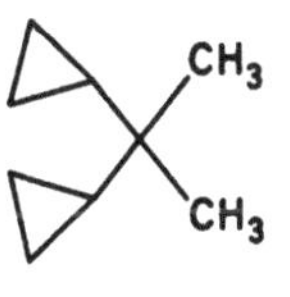


257 (84%)

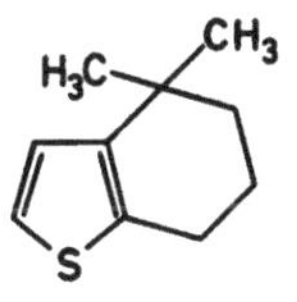


258 (74%)

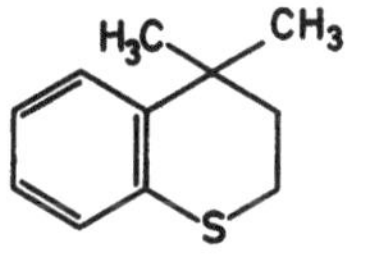


259 (77%)

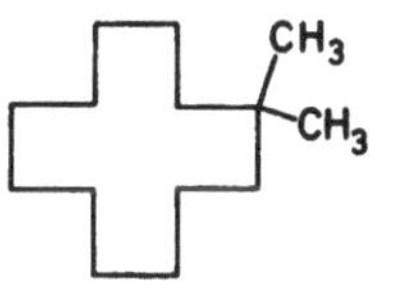


260 (73%)

## F.III Direct Geminal Dialkylation of Ketones

Direct geminal dialkylation of ketones using homologous compounds of the type $R_2TiCl_2$ does not proceed as smoothly. Also, the introduction of two different R groups cannot be accomplished in an analogous manner. Besides the three step sequence outlined in Section F.I, we have developed a one-pot procedure. Addition of alkyllithium to ketones *242* in hexane followed by the addition of a mixture of *17* and *171* (prepared by quenching 1.5 equiv. $(CH_3)_2Zn$ and 2.0 equiv. $TiCl_4$ in $CH_2Cl_2$) at —40 to —10 °C affords the mixed dialkyl product *262* in good yield (Equation 83) [133]. This unusual reaction requires more active methyl groups than the procedure involving tert-alkyl chlorides (Sect. F.I), but is simpler. Furthermore, it is rather useful if the corresponding tert-alkyl chlorides are sensitive and/or difficult to prepare.

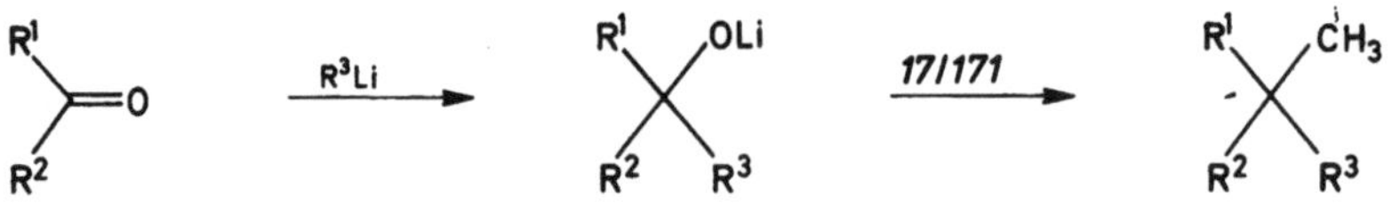


*242* *261* *262* (83)

An interesting example of an application of this method pertains to the synthesis of pharmacologically active synthetic $\Delta^1$-tetrahydrocannabinoids [134] of the type *267* which have a lipophilic tertiary alkyl side-chain. Equation 84 shows that organotitanium chemistry provides a versatile means to prepare the precursors *264* (65–80%) [133]. Demethylation of *264* using trimethylsilyl chloride and sodium iodide affords the resorcinol derivatives *265* (~95%) [133]. Compounds of this type have been previously condensed with *266* in the presence of acids to form the $\Delta^{1(6)}$-isomers of *267*, which in turn can be converted into *267* [135]. It should be mentioned that the meta-substitution pattern of *265* prohibits simple Friedel-Crafts alkylation of resorcinol, which is the reason why alternative multistep syntheses of *264* have had to be developed [134–136].

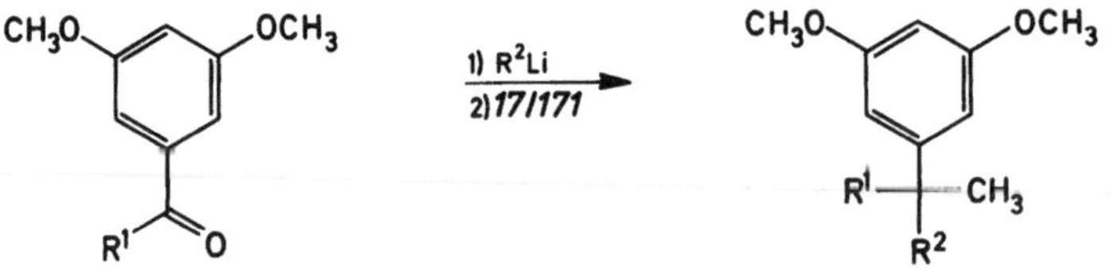


*263* *264* (84)

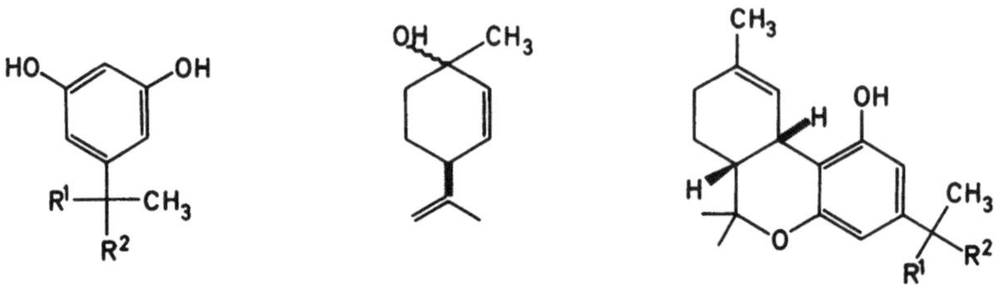


*265* *266* *267*

The above C—C bond forming reaction is closely related to methylation of tertiary alcohols and diols [128], gem-dichlorides [137], ethers [137] and ketals [35]. It is also possible to convert acid chlorides directly into tert-butyl derivatives (85% conversion) (Equation 85) [35].

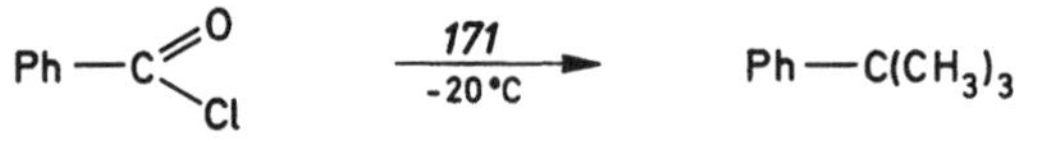


*268* *269* (85)

## F.IV Methylation of Secondary Alkyl Halides

Relative to tertiary alkyl halides, secondary derivatives react considerably slower. At room temperature and long reaction periods (~24 h) cyclohexyl chloride is almost quantitatively methylated with dimethyltitanium dichloride (prepared in situ from dimethylzinc and catalytic amounts of $TiCl_4$) [137], but other cyclic or acyclic halides tend to undergo competing rearrangements prior to C—C bond formation [77]. The same applies to 1,2-dihalides such as 1,2-dibromocyclohexane which affords 1,1-dimethylcyclohexane instead of the 1,2-dimethyl derivative [137]. In complete contrast, activated secondary chlorides behave much like tertiary derivatives, i.e., methylation is fast and position specific at low temperatures. Examples are shown in Equation 86 [137]. It should be noted that in such cases cuprate chemistry affords less than 40% of methylation products [138].

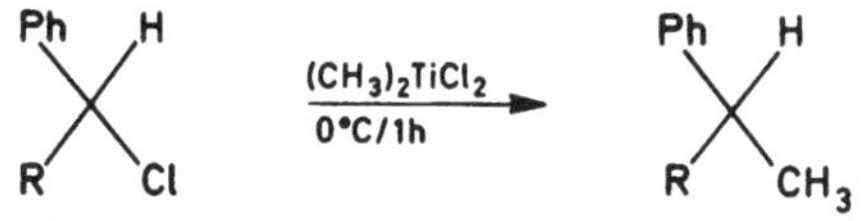


270 271 (86)

R = phenyl; 95% R = methyl; 90% R = n-butyl; 78%

In summary, methylation of secondary alkyl halides is essentially restricted to $S_N1$ active derivatives. In all cases Lewis acidic titanium species must be used (Sect. B.I). Methyltitanium triisopropoxide is much less Lewis acidic and does not participate in the above substitution reactions [128]. All of these observations are in line with $S_N1$ ionization of the alkyl halide followed by C—C bond formation. $S_N2$ reactions of organotitanium reagents with methyl iodide have not been explored systematically, but appear to be slow. For example, titanium enolates as described in Section D.III.1 do not react at room temperature with methyl iodide or allylbromide [77]. The reaction of allyltitanium species with alkyl halides is also sluggish [77]. Acylation using acid chlorides is very fast, however [25]. The introduction of hetero-electrophiles is possible for most organotitanium reagents. Halogenation using $Br_2$ or $I_2$ or hydroxylation employing $O_2$ is fast and nearly quantitative, e.g., in reactions of titanium enolates or aryltitanium reagents, respectively [77].

## F.V Wittig-type Olefination

The so-called Tebbe reagent *272* reacts with carbonyl compounds to form olefinic products [139a], a reaction which is of particular synthetic value in case of esters (Equation 87) [139b].

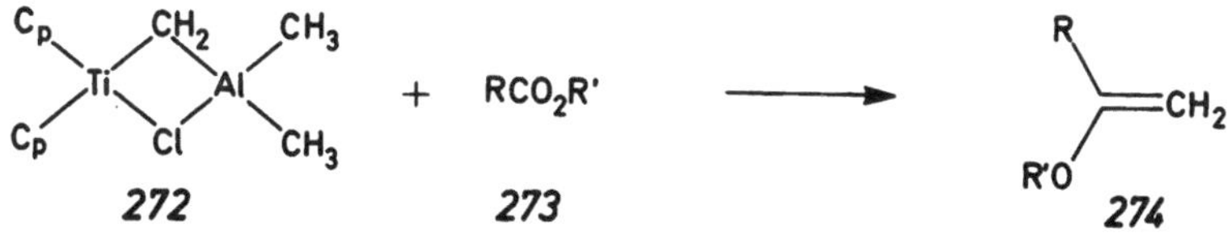


272 273 274 (87)

Methylenation of ketones and aldehydes can also be carried out using the combination $CH_2X_2$ (X = Br, I)/Zn/$TiCl_4$ according to Nozaki[140], a procedure which may involve intermediate organotitanium species. It is very useful in a number of cases in which the classical Wittig reaction fails. An excess of compounds of the type *275* reacts chemoselectively with aldehydes in the presence of ketones to form the corresponding terminal olefins directly as reported by Kauffmann [141] (Equation 88). The corresponding triisopropoxy compound affords the addition products without undergoing subsequent Peterson elimination [77]. However, the adducts can be treated with KH to afford the olefins [77]. Finally, mixed metal reagents of the type *226* (Sect. D.III.3) allow for stereoselective diene syntheses [90].

$$\underset{\textit{275}}{(CH_3)_3SiCH_2TiCl_3} \xrightarrow{RCHO} \underset{\textit{276}}{RCH{=}CH_2} \qquad (88)$$

## G. A Few Hints on how to use Organotitanium Reagents

Other than the restrictions already mentioned, several additional aspects should be kept in mind when applying organotitanium chemistry as presented here. In using reaction types discussed in Section F.I–F.IV, workup is best performed by quenching with ice water, which means slightly acidic conditions. If $Na_2CO_3$ is used, $TiO_2$ containing emulsions may occur which hamper workup. In the more extensive area of carbanion selectivity (Sect. C–E), the reaction mixtures are poured on dilute HCl solutions. If this causes emulsionformation, a saturated solution of $NH_4F$ or KF is the method of choice. In such cases adjustment of the pH to about 6 using acids may be beneficial.

If carbanions are to be titanated, alkoxy or amino ligands at titanium are most likely to ensure success. Sulfur or phosphorus ligands have not been tested. In rare cases electrontransfer instead of titanation sets in, forming Ti(III) species which are generally unsuitable for useful chemistry. This is most likely to occur if the carbanion is very electron rich, e.g., in case of dianions or extended anionic π-systems. We have noticed that this undesired property *decreases* in going from chloro to alkoxy and finally to amino ligands at titanium. For example, dianions derived from carboxylic acids reduce chlorotitanium triisopropoxide *3* to some extent, whereas quantitative double titanation occurs with chloro- or bromotitanium tris(diethylamide) *15* [77]. Addition of amines to the reaction mixture has similar effects [77].

## H. Conclusions and Further Work

This review describes some novel reaction types (Section F), but the more extensive area of activity has to do with variable adjustment of carbanion reactivity and selectivity in reactions with carbonyl compounds (Sects. C–E). Organotitanium chemistry seems to be complementary to that of Li, Mg, Zn, Cu, Fe, Ni and Pd. For example, diastereoselectivity is frequently better or opposite to that obtained by using conventional lithiated species. Although the vast number of "carbanions" have not been titanated, the basic principles are clear. Enhancement of chemoselectivity by prior titanation appears to be a simple goal. Controlling stereochemistry is by nature more difficult. In the synthesis of particular target molecules, optimization is possible via variation of the ligands at titanium. Finally, it is likely that additional reaction types involving titanium species will emerge, e.g., stereoselective Michael additions.

During the short time between submission and correction of this review further advances in organotitanium chemistry have been made. Allyltitanium tris(diethylamide) *102* reacts smoothly with various N-acyl-imidazoles to form β,γ-unsaturated ketones [142]. The allyltitanium ate complex *99* attacks styrene oxide regioselectively at the carbon atom bearing the phenyl group (~96% conversion), in contrast to the corresponding Grignard reagent *96* which leads to a mixture of regioisomers [142]. This is the first example in which an organotitanium reagent reacts whith an epoxide. Crotyltitanium triphenoxide is an excellent reagent for threo-selective addition to aldehydes [143], surpassing the analogous triisopropoxy derivative *220* [90]. Crotyl derivatives such as crotyltitanium tris(diethylamide) *219* react threo-selectively even with unsymmetrical *ketones* [142], e.g., Equation 89.

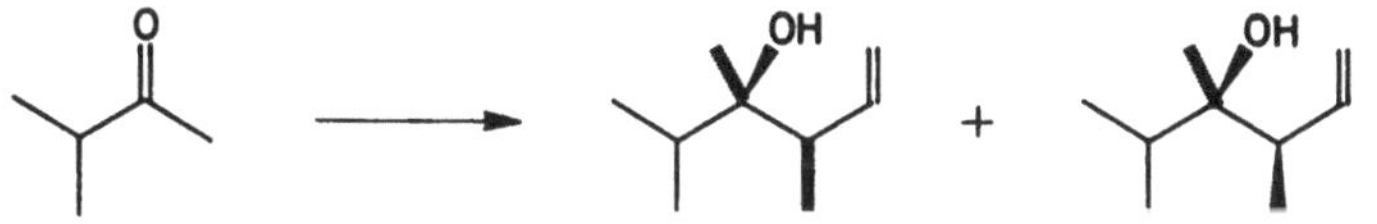

*277* → *278* + *279* (89)

| | 278 | 279 |
|---|---|---|
| *219* | 97 | 3 |
| *113* | 52 | 48 |

The intriguing behavior of allyl- and crotyltitanium ate complexes with regard to controlled aldehyde or ketone selectivity as described in Section C.VII is general. Thus, methallyl reagents display the same sort of reactivity pattern; switching the ligands at titanium allows for reversal of chemoselectivity. In a competition experiment using 2-heptanone and diisopropyl ketone, the alkoxy ate complex *99* picks out only the former, while the amino ate complex *106* favors the latter which is sterically *more* hindered (63:37 product ratio)! Progress in 1,2-asymmetric induction using organotitanium reagents has also been recorded. For example, methyltitanium triisopropoxide *6* reacts with 2-benzyloxypropanal (O-benzyllactaldehyde) to yield a 90:10 mixture of Cram and anti-Cram addition products (>90% yield) [144]. Interestingly, the more Lewis acidic methyltitanium trichloride *17* reverses diastereoselectivity (8:92 product ratio) [144]. Although other ligands at titanium have not yet been tested, the results again demonstrate the principle of variable adjustment of carbanion selectivity by conversion into titanium reagents [21]. Finally, α-deprotonated acyl silanes [108] can be titanated with chlorotitanium triisophenoxide to form the corresponding titanium enolates which react erythro-selectively with aldehydes [145]. For example, the mixture of Z- and E-enolates derived from ethyl trimethylsilyl ketone adds to benzaldehyde to afford a 92:8 mixture of erythro and threo adducts.

Perhaps the most interesting new development concerns enantioselective addition reactions involving reagents with a *center of chirality at titanium*. The extreme difficulties in this endeavor have been delineated in Section E. Very recent work at Marburg points to one of several possible solutions to this problem [146]. Titanium compounds bearing two different $h^5$-cyclopentadienyl ligands as well as two different σ-bonded groups can be separated into antipodes which are known to be configurationally stable [147], e.g., *280*. Unfortunately, replacing the chloro ligand by alkyl groups leads to compounds which are not reactive towards aldehydes [146]. In sharp contrast, it turned out that allyl derivatives are in fact reactive enough. In a preliminary series of experiments we prepared optically active *280* [147], quenched it with allylmagnesium chloride *96* and added benzaldehyde in a one-pot procedure. In this non-optimized reaction the product *111* was in fact optically active (ee = 11%) [146]. Several important aspects are being studied currently. Does the $MgCl_2$ which is present affect enantioselectivity? Does the substitution step *280→281*[24] occur with retention or inversion of configuration, or perhaps partial racemization? The above initial ee-value is low. However, the reaction represents the first case of an enantioselective Grignard-type addition in which the metal atom is the chiral center [148].

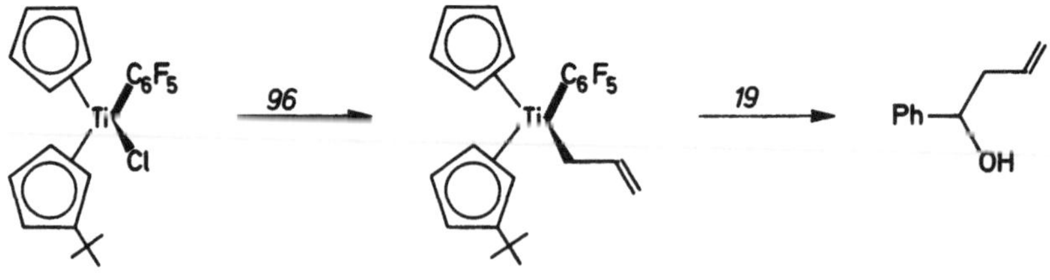


(90)

*280* *281* *111*

This work was supported by the Deutsche Forschungsgemeinschaft and the Fond der Chemischen Industrie.

## I. References


1. a) Segnitz, A.: in Houben-Weyl-Müller, Methoden der Organischen Chemie, Vol. 13/7, p. 263, Stuttgart, Thieme Verlag 1975; b) Gmelin Handb., Titanorganische Verbindungen, Vol. 40, N.Y. 1977
2. a) Sinn, H., Kaminsky, W.: Adv. Organomet. Chem. *18*, 99 (1980); b) Pino, P., Mülhaupt, R.: Angew. Chem. *92*, 869 (1980); Angew. Chem. Int. Ed. Engl. *19*, 857 (1980)
3. Chatt, J., Delworth, J. R., Richards, R. L.: Chem. Rev. *78*, 589 (1978)
4. See for example: a) Fachinetti, G., Floriani, C.: J. Organomet. Chem. *71*, C5 (1974); b) McDermott, J. X., Whitesides, G. M.: J. Am. Chem. Soc. *96*, 947 (1974); c) Fachinetti, G. et al.: J. Chem. Soc., Dalton Trans. *1977*, 2297; d) Shoer, L. J., Schwartz, J.: J. Am. Chem. Soc. *99*, 5832 (1977); e) Klei, B., Teuber, J. H., de Liefde Meijer, H. J.: Chem. Comm. *1981*, 342


[24] We have arbitrarily pictured the substitution step as occuring with retention of configuration.

5. Dormond, A. et al.: J. Organomet. Chem. *177*, 181 (1979)
6. Lee, J. B. et al.: J. Am. Chem. Soc. *103*, 7358 (1981); and lit. cited therein
7. See for example: a) Kano, S. et al.: Synthesis *1980*, 695, 741; b) Colomer, E., Corriu, R.: J. Organomet. Chem. *21*, 381 (1970); c) Neese, H. J., Bürger, H.: J. Organomet. Chem. *32*, 213
8. a) Schwartz, J.: Angew. Chem. *88*, 402 (1976); Angew. Chem., Int. Ed. Engl. *15*, 333 (1976); b) Ashby, E. C., Noding, S. A.: J. Org. Chem. *45*, 1035 (1980); c) Sato, F. et al.: Chem. Lett. *1980*, 99; d) Coleman, R. A. et al.: J. Organomet. Chem. *146*, 221 (1978); e) Eisch, J. J., Galle, J. E.: J. Organomet. Chem. *160*, C8 (1978); f) Carr, D. B., Schwartz, J.: J. Am. Chem. Soc. *101*, 3521 (1979); g) Bogdanovic, B. et al.: Angew. Chem. *94*, 206 (1982); Angew. Chem., Int. Ed. Engl. *21*, 199 (1982); Angew. Chem. Suppl. *1982*, 457–470; h) Fell, B., Asinger, F., Sulzbach, R. A.: Chem. Ber. *103*, 3830 (1970)
9. Review: a) Negishi, E.: Pure Appl. Chem. *53*, 2333 (1981); see also: b) Brown, D. C. et al.: J. Org. Chem. *44*, 3457 (1979); c) Schiavelli, M. D., et al.: J. Org. Chem. *46*, 807 (1981); d) Wilke, G.: J. Organomet. Chem. *200*, 349 (1980); e) Yoshida, T.: Chem. Lett. *1982*, 293
10. Pez, G. P., Armor, J. N.: Adv. Organomet. Chem. *19*, 2 (1981)
11. a) McMurry, J. E.: Acc. Chem. Res. *7*, 281 (1974); b) review of reductions using other low-valent Ti compounds: Castedo, L. et al.: J. Org. Chem. *46*, 4292 (1981) and lit. cited therein; c) Rieke, R. D.: Acc. Chem. Res. *10*, 301 (1977)
12. Mukaiyama, T.: Angew. Chem. *89*, 858 (1977); Angew. Chem., Int. Ed. Engl. *16*, 817 (1977)
13. Reetz, M. T.: Angew. Chem. *94*, 97 (1982); Angew. Chem., Int. Ed. Engl. *21*, 96 (1982)
14. Chan, T. H., Fleming, I.: Synthesis *1979*, 761
15. See for example: Chem. Abstr. *61*: 4223f (1964); *67*: 73173b (1967); *74*: 100452v (1971)
16. Seebach, D. et al.: Synthesis *1982*, 138
17. „Titansäureester", Information Pamphlet, Dynamit Nobel AG, Troisdorf, W-Germany
18. Katsuki, T., Sharpless, K. B.: J. Am. Chem. Soc. *102*, 5974 (1980)
19. a) Labinger, J. A.: J. Organomet. Chem. *227*, 341 (1982); b) Labinger, J. A.: J. Organomet. Chem. *196*, 37 (1980); c) Labinger, J. A.: J. Organomet. Chem. *180*, 187 (1979); d) Labinger, J. A.: J. Organomet. Chem. *167*, 19 (1979); e) Clark, R. J. H., Moorhouse, S., Stockwell, J. A.: Organomet. Chem. Rev., Vol. III, p. 223, ((ed.) Seyferth, D.), 1977; f) Clark, R. J. H.: Comprehensive Inorganic Chem., Vol. III, p. 355, Oxford, Pergamon Press 1973; g) Feld, G., Cowe, P. L.: The Organic Chemistry of Titanium, London, Butterworths 1965; h) Clark, R. J. H.: The Chemistry of Titanium and Vanadium, Amsterdam, Elsevier 1968; i) Kepert, D. L.: The Early Transition Metals, N.Y., Academic Press 1972; j) Wailes, P. C., Coutts, R. S. P., Weigold, H.: Organometallic Chemistry of Titanium, Zirconium and Hafnium, N.Y., Academic Press 1974
20. Stowell, J. C.: Carbanions in Organic Synthesis, N.Y., Wiley 1979
21. a) Reetz, M. T. et al.: Chem. Industry *1981*, 541; b) Wenderoth, B.: Diplomarbeit, Univ. Marburg *1980*
22. Reetz, M. T.: Nachr. Chem. Lab. Techn. *29*, 165 (1981)
23. Dijkgraaf, C., Rousseau, J. P. G.: Spectrochim. Acta, A, *24*, 1213 (1968)
24. Westermann, J.: projected Dissertation, Univ. Marburg
25. Reetz, M. T., Peter, R.: Tetrahedron Lett. *1981*, 4691
26. Wilkinson, G.: Pure Appl. Chem. *30*, 627 (1972)
27. Davidson, P. J., Lappert, M. F., Pearce, R.: Chem. Rev. *76*, 219 (1976)
28. Schrock, R. R., Parshall, G. W.: Chem. Rev. *76*, 243 (1976)
29. a) Baird, M. C.: J. Organomet. Chem. *64*, 289 (1974); b) Braterman, P. S.: Top. Curr. Chem. *92*, 149 (1980); c) Akermark, B., Ljungqvist, A. J.: Organomet. Chem. *182*, 59 (1979)
30. a) Herman, D. F., Nelson, W. K.: J. Am. Chem. Soc. *74*, 2693 (1952); b) Herman, D. F., Nelson, W. K.: J. Am. Chem. Soc. *75*, 3877 (1953); c) Herman, D. F., Nelson, W. K.: J. Am. Chem. Soc. *75*, 3882 (1953)
31. a) Beermann, C., Bestian, H.: Angew. Chem. *71*, 618 (1959); b) Clauss, K.: Liebigs Ann. Chem. *711*, 19 (1968); c) Clauss, K., Bestian, H.: Liebigs Ann. Chem. *654*, 8 (1962)
32. Rausch, M. D., Gordon, H. B.: J. Organomet. Chem. *74*, 85 (1974)
33. Kühlein, K., Clauss, K.: Makromol. Chem. *155*, 145 (1972)
34. a) Sugahara, H., Shuto, Y.: J. Organomet. Chem. *24*, 709 (1970); b) Yoshino, A., Shuto, Y., Iitaka, Y.: Acta Crystall., Sect. B, *26*, 744 (1970)
35. Reetz, M. T., Westermann, J.: unpublished results 1979–80, Univ. Bonn
36. Roberts, R. M. G.: J. Organomet. Chem. *63*, 159 (1973)

37. Collier, M. R., Lappert, M. F., Pearce, R.: J. Chem. Soc., Dalton Trans. *1973*, 445
38. a) Clauss, K., Beermann, C.: Angew. Chem. *71*, 627 (1959); b) Berthold, H. J., Groh, G.: Z. Anorg. Chem. *319*, 230 (1963); c) Müller, J., Thiele, K. H.: Z. Anorg. Chem. *362*, 120 (1968);
39. a) Bürger, H., Neese, H. J.: J. Organomet. Chem. *20*, 129 (1969); b) Bürger, H., Neese, H. J.: J. Organomet. Chem. *21*, 381 (1970); c) Neese, H. J., Bürger, H.: J. Organomet. Chem. *32*, 213 (1971); d) Bürger, H., Neese, H. J.: J. Organomet. Chem. *36*, 101 (1972)
40. Thiele, K. H. et al.: Z. Anorg. Chem. *378*, 62 (1970)
41. Reetz, M. T., Westermann, J., Steinbach, R.: Angew. Chem. *92*, 931 (1980); Angew. Chem., Int. Ed. Engl. *19*, 900 (1980)
42. Gray, A. P., Callear, A. B., Edgecombe, F. H. C.: Can. J. Chem. *41*, 1502 (1963)
43. Gorisch, R. D.: J. Am. Chem. Soc. *82*, 4211 (1960)
44. Thiele, K. H.: Pure Appl. Chem. *30*, 575 (1972)
45. Dong, D.: Inorg. Chim. Acta *29*, L225 (1978)
46. a) Bassi, I. W. et al.: J. Am. Chem. Soc. *93*, 3787 (1971); b) Davies, G. R., Jarvis, J. A. J., Kilbourn, B. T.: Chem. Comm. *1971*, 1511
47. Davies, G. R. et al.: Chem. Comm. *1971*, 677
48. Evans, D. A., Nelson, J. V., Taber, T. R.: Topics in Stereochem., *13*, 1 (1982)
49. Telnoi, V. I. et al.: Dokl. Akad. Nauk (SSSR) *174*, 1374 (1967)
50. a) Lappert, M. F., Patil, D. S., Pedley, J. B.: Chem. Comm. *1975*, 830; b) review: Connor, J. A.: Top. Curr. Chem. *71*, 71 (1977)
51. Cox, J. D., Pilcher, G.: Thermochemistry of Organic and Organometallic Compounds, London, Academic Press 1970
52. Dijkgraaf, C., Rousseau, J. P. G.: Spectrochim. Acta, *A*, 25, 1455 (1969)
53. See ref. 1b, p. 19
54. Armstrong, D. R., Perkins, P. G., Stewart, J. J. P.: Rev. Roumaine Chim. *20*, 177 (1975)
55. Hotokka, M., Pyykkö, P.: J. Organomet. Chem. *174*, 289 (1979)
56. Blustin, P. H.: J. Organomet. Chem. *210*, 357 (1981)
57. a) Graham, G. D., Marynick, D. S., Lipscomb, W. N.: J. Am. Chem. Soc. *102*, 4572 (1980): b) Jemmis, E. D., Chandrasekhar, J., Schleyer, P. v. R.: J. Am. Chem. Soc. *101*, 2848 (1979); c) Streitwieser, A. et al.: J. Am. Chem. Soc. *98*, 4778 (1976)
58. a) Cassoux, P., Crasnier, F., Laberre, J. F.: J. Organomet. Chem. *165*, 303 (1979); b) McKinney, R. J.: Chem. Comm. *1980*, 490; and lit. cited therein
59. Lappert, M. F., Pedley, J. B., Sharp, G.: J. Organomet. Chem. *66*, 271 (1974)
60. Basso-Bert, M. et al.: J. Organomet. Chem. *136*, 201 (1977)
61. a) Caulton, K. G. et al.: J. Am. Chem. Soc. *102*, 3009 (1980); b) Caulton, K. G. et al.: J. Organomet. Chem. *201*, 389 (1980)
62. a) Blandy, C., Guerreiro, M. R., Gervais, D.: C. R. Acad. Sc., Serie C, *278*, 1323 (1974); b) McAlees, A. J., McCrindle, R., Moon-Fat, A. R.: Inorg. Chem. *15*, 1065 (1976); and lit. cited therein
63. Ireland, R.: Organic Synthesis, Englewood Cliffs, N.J., Prentice Hall 1969
64. a) Nützel, K.: in Houben-Weyl-Müller, Methoden der Organischen Chemie, Vol. 13/2a, p. 47, Stuttgart, Thieme-Verlag 1973; b) Schöllkopf, U.: in Houben-Weyl-Müller, Methoden der Organischen Chemie, Vol. 13/1, p. 87, Stuttgart 1970
65. a) Kharasch, M. S., Cooper, J. H.: J. Org. Chem. *10*, 46 (1945); b) Vaskan, R. N. et al.: Zh. Org. Khim. *11*, 1818 (1975)
66. Reetz, M. T., Westermann, J.: unpublished results, Univ. Bonn 1980
67. Nützel, K.: in Houben-Weyl-Müller, Methoden der Organischen Chemie, Vol. 13/2a, p. 553, Stuttgart, Thieme-Verlag 1973
68. Nützel, K.: in Houben-Weyl-Müller, Methoden der Organischen Chemie, Vol. 13/2, p. 859, Stuttgart, Thieme-Verlag 1973
69. Yamamoto, Y. et al.: J. Org. Chem. *47*, 119 (1982)
70. a) Cahiez, G., Bernard, D., Normant, J. F.: Synthesis *1977*, 130; b) Kauffmann, T., Hamsen, A., Beirich, C.: Angew. Chem. *94*; 145 (1982); Angew. Chem., Int. Ed. Engl. *21*, 144 (1982)
71. The early part of our work was reported in lectures presented at Univ. München, (November 10, 1979) and Bayer AG, Krefeld (November 15, 1979)
72. Reetz, M. T. et al.: Angew. Chem. *92*, 1044 (1980); Angew. Chem., Int. Ed. Engl. *19*, 1011 (1980)
73. Peter, R.: Diplomarbeit, Univ. Marburg 1981

74. Reetz, M. T., Westermann, J., Steinbach, R.: Angew. Chem. *92*, 933 (1980); Angew. Chem., Int. Ed. Engl. *19*, 901 (1980)
75. Reetz, M. T.: Schriftenreihe 30 Jahre Fonds der Chemischen Ind., Frankfurt 1980
76. Weidmann, B., Seebach, D.: Helv. Chim. Acta *63*, 2451 (1980)
77. Reetz, M. T. et al.: unpublished results, Univ. Marburg 1980–82
78. Weidmann, B. et al.: Helv. Chim. Acta *64*, 357 (1981)
79. Reetz, M. T., Maus, S.: in preparation
80. Weidmann, B., Maycock, C. D., Seebach, D.: Helv. Chim. Acta *64*, 1552 (1981)
81. $TiCl_4$ and $Ti(OR)_4$ are cheap raw materials used in large quantities in industry. Zirconium analogs cost about 3–4 times as much
82. Reetz, M. T. et al.: submitted for publication
83. Reetz, M. T. et al.: Angew. Chem. *94*, 133 (1982); Angew. Chem., Int. Ed. Engl. *21*, 135 (1982); Angew. Chem. Suppl. *1982*, 257–268
84. See Ref. 13, footnote 65
85. Kostova, K. et al.: Helv. Chim. Acta *65*, 249 (1982)
86. Schlosser, M.: Struktur und Reaktivität polarer Organometalle, Berlin, Springer-Verlag 1973
87. Wakefield, B. J.: The Chemistry of Organolithium Compounds, N.Y., Pergamon Press 1974
88. a) Pinkus, A. G., Servoss, W. C.: J. Chem. Soc. Perkin II *1979*, 1600; b) Levina, R. Y. et al.: Zh. Obshch. Kim. *32*, 1377 (1962)
89. Berthold, H. J., Groh, G.: Angew. Chem. *78*, 495 (1966); Angew. Chem., Int. Ed. Engl. *5*, 516 (1966)
90. Reetz, M. T., Wenderoth, B., Stninbach, R.: submitted for publication
91. Review: Hoffmann, R. W.: Angew. Chem., *94*, 569 (1982)
92. Okude, Y. et al.: J. Am. Chem. Soc. *99*, 3179 (1977)
93. Naruta, Y., Ushida, S., Maruyama, K.: Chem. Lett. *1979*, 919
94. a) Cram, D. J., Elhafez, F. A. A.: J. Am. Chem. Soc. *74*, 5828 (1952); b) Cram. D. J., Allinger, J.: J. Am. Chem. Soc. *76*, 4516 (1954); c) Cram, D. J., Kopecky, K. R.: J. Am. Chem. Soc. *81*, 2748 (1959)
95. a) Morrison, J. D., Mosher, H. S.: Asymmetric Organic Reactions, Englewood Cliffs, N. J., Prentice Hall 1971; b) Anh, N. T.: Top. Curr. Chem. *88*, 40 (1980); c) Jones, P. R., Goller, E. J., Kauffman, W. J.: J. Org. Chem. *36*, 3311 (1971)
Kauffmann, W. J.: J. Org. Chem. *36*, 3311 (1971)
96. Clark, R. J. H., McAlees, A. J.: Inorg. Chem. *11*, 342 (1972)
97. Reetz, M. T., Westermann, J.: Synth. Comm. *11*, 647 (1981)
98. Makino, T.: J. Org. Chem. *43*, 276 (1978)
99. Piatak, D. M., Wicha, J.: Chem. Rev. *78*, 199 (3978)
100. Stocker, J. H. at al.: J. Am. Chem. Soc. *82*, 3913 (1960)
101. Marsella, J. A., Moloy, K. G., Caulton, K. G.: J. Organomet. Chem. *201*, 389 (1980)
102. Reetz, M. T., Steinbach, R.: unpublished results, Univ. Marburg 1982
103. a) Wolfrom, M. L., Hanessian, S.: J. Org. Chem. *27*, 1800 (1962); b) Inch. T. D.: Carbohydrate Res. *5*, 45 (1967)
104. a) Still, W. C., Schneider, J. A.: Tetrahedron Lett. *1980* 1035; b) Narasaka, K.: Chem. Lett. *1980*, 1415
105. Ashby, E. C., Laemmle, J. T.: Chem. Rev. *75*, 521 (1975)
106. Ashby, E. C., Noding, S. A.: J. Org. Chem. *44*, 4371 (1979)
107. MacDonald, T. L., Still, W. C.: J. Am. Chem. Soc. *97*, 5280 (1975)
108. Heathcock, C. H.: Comprehensive Carbanion Chemistry, Vol. II, ((ed.) Durst, T., Buncel, E.), Amsterdam, Elsevier 1981
109. a) Evans, D. A., McGee, L. R.: Tetrahedron Lett. *1980*, 3975; b) Yamamoto, Y., Maruyama, K.: Tetrahedron Lett. *1980*, 4607
110. Yamamoto, Y., Yatagai, H., Maruyama, K.: Chem. Comm. *1981*, 162
111. Noyori, R., Nishida, I., Sakata, J.: J. Am. Chem. Soc. *103*, 2106 (1981)
112. Evans, D. A., McGee, L. R.: J. Am. Chem. Soc. *103*, 2876 (1981)
113. Stork, G., Shiner, C. S., Winkler, J. D.: J. Am. Chem. Soc. *104*, 310 (1982)
114. Posner, G. H.: Pure Appl. Chem. *53*, 2307 (1981)
115. Reetz, M. T., Steinbach, R., Keßeler, K.: Angew. Chem., in press
116. a) Corey, E. J., Enders, D.: Chem. Ber. *111*, 1337 (1978); b) Davenport, K. G. et al.: J. Am. Chem. Soc. *101*, 5654 (1979)

117. Hoffmann, R. W. et al.: J. Org. Chem. *46*, 1309 (1981)
118. Yamamoto, Y. et al.: J. Am. Chem. Soc. *102*, 7107 (1980)
119. Sato, F., Iijima, S., Sato, M.: Tetrahedron Lett. *1981*, 243
120. Sato, F. et al.: Chem. comm. *1981*, 1140
121. Yamamoto, Y., Maruyama, K.: Tetrahedron Lett. *1981*, 2895
122. Tsai, D. J. S., Matteson, D. S.: Tetrahedron Lett. *1981*, 2751
123. Hanko, R., Hoppe, D.: Angew. Chem., *94*, 378 (1982); Angew. Chem., Int. Ed. Engl. *21*, 372 (1982)
124. Olivero, A. G., Weidmann, B., Seebach, D.: Helv. Chim. Acta *64*, 2485 (1981)
125. Trost, B. M.: Acc. Chem. Res. *7*, 85 (1974)
126. a) Noller, C. R.: J. Am. Chem. Soc. *51*, 594 (1929); b) Buck, R. F.: J. Inst. Petrol. *34*, 339 (1948)
127. a) Kennedy, J. P., Desai, U. V., Sivaram, S.: J. Am. Chem. Soc. *95*, 6386 (1973); b) see however: Beckhaus, H. D., Hellmann, G., Rüchardt, C.: Chem. Ber. *111*, 72 (1978)
128. Reetz, M. T. et al.: Chem. Comm. *1980*, 1202
129. Reetz, M. T., Stephan, W.: J. Chem. Res. *1981*, (S) 44; J. Chem. Res. *1981*, (M) 0583-0594
130. Reetz, M. T., Westermann, J., Steinbach, R.: Chem. Comm. *1981*, 237
131. Meisters, A., Mole, T.: Aust. J. Chem. *27*, 1665 (1974)
132. a) Oppolzer, W., Godel, T.: J. Am. Chem. Soc. *100*, 2583 (1978); b) Gröger, C., Musso, H., Rossnagel, I.: Chem. Ber. *113*, 3621 (1980); c) Martin, S. F.: Tetrahedron *36*, 419 (1980); d) Trost, B. M., Hiemstra, H.: J. Am. Chem. Soc. *104*, 886 (1982); e) Pataki, J., Konieczny, M., Harvey, R. G.: J. Org. Chem. *47*, 1133 (1982)
133. Reetz, M. T., Westermann, J.: submitted
134. Review of cannabinoids: a) Mechoulam, R., McCallum, N. K., Burstein, S.: Chem. Rev. *76*, 75 (1976); b) Mechoulam, R., Carlini, E. A.: Naturwiss. *65*, 174 (1978)
135. Petrzilka, T., Haefliger, W., Sikemeier, C.: Helv. Chim. Acta *52*, 1102 (1969)
136. Singh, V., Kaul, V. V., Martin, A. R.: Synth. Comm. *1981*, 429
137. Reetz, M. T., Steinbach, R., Wenderoth, B.: Synth. Comm. *1981*, 261
138. Posner, G. H., Brunelle, D. J.: Tetrahedron Lett. *1972*, 293
139. a) Tebbe, F. N., Parshall, G. W., Reddy, G. S.: J. Am. Chem. Soc. *100*, 3611 (1978); b) Pine, S. H. et al.: J. Am. Chem. Soc. *102*, 3270 (1980)
140. a) Takai, K. et al.: Tetrahedron Lett. *1978*, 2417; b) Takai, K. et al.: Bull. Chem. Soc. Jap. 53, 1698 (1980)
141. Kauffmann, T. et al.: Tetrahedron Lett. *1981*, 5031
142. Reetz, M. T., Wenderoth, B.: unpublished results
143. Widler, L., Seebach, D.: Helv. Chim. Acta *65*, 1085 (1982)
144. Reetz, M. T., Schmidtberger, S.: in preparation
145. Reetz, M. T., Peter, R.: unpublished results
146. Reetz, M. T., Westermann, J.: in preparation
147. Leblanc, J. C., Moise, C., Tirouflet, J.: Nouv. Chim. *1*, 211 (1977)
148. For an excellent review concerning chiral tetrahedral organotransition metal compounds, see Brunner, H.: Adv. Organomet. Chem. *18*, 151 (1980)

# Lithium Halocarbenoids — Carbanions of High Synthetic Versatility

*Dedicated to Prof. Georg Wittig on the occasion of his 85th birthday*

**Herbert Siegel**[1]

Laboratorium für Organische Chemie der Eidgenössischen Technischen Hochschule, ETH Zentrum, Universitätstraße 16, CH-8092 Zürich, Switzerland
Present address: Hauptlaboratorium der Hoechst AG, Postfach 800320, D-6230 Frankfurt 80, FRG

## Table of Contents

**A. Introduction** . . . . . . . . . . . . . . . . . . . . . . . . . 56

**B. Preparative Reactions of Lithium Halocarbenoids** . . . . . . . . . . . . 57
B.I Carbon Atom Insertions . . . . . . . . . . . . . . . . . . . . 57
B.II Reactions of Lithium Carbenoids with Electrophiles . . . . . . . . 59
B.II.1 Alkylidene Lithium Halocarbenoids . . . . . . . . . . . . 59
B.II.2 α-Heteroatom-Substituted Lithium Halocarbenoids . . . . . . 63
B.II.3 Cyclopropylidene Lithium Halocarbenoids . . . . . . . . . . 68
B.II.4 Vinylidene and Allylidene Lithium Halocarbenoids . . . . . . 71

**C. Conclusion** . . . . . . . . . . . . . . . . . . . . . . . . . 75

**D. References** . . . . . . . . . . . . . . . . . . . . . . . . . 76

Lithium halocarbenoids are no more subject of mechanistic interest only. Improvement of preparative techniques in the last ten years made them to valuable synthetic intermediates which are stable in the temperature range between —130 and —70 °C. They are generated from readily available starting materials and give high yields of adducts on reaction with electrophiles.

The synthetic applications of halocarbenoids are mainly determined by the framework bearing the carbenoid center. This article describes the different kinds of synthetic transformations that can be achieved by the use of alkylidene, α-heterosubstituted, cyclopropylidene, vinylidene, and allylidene lithium halocarbenoids. Their particuliar value in organic synthesis results from various rearrangement reactions of the primary adducts formed by reaction of the carbenoid with the electrophile.

[1] The author thanks Professor D. Seebach for his encouragement and help in writing this review.

# A. Introduction

Organolithium derivatives containing α-heterosubstituents derived from elements of the fourth, fifth and sixth main group (Si; N, P, As; O, S, Se) can be used for a multitude of C—C bond forming processes and have been shown to possess considerable synthetic potential [1]. The corresponding intermediates bearing elements of the seventh main group (F, Cl. Br, I) at the metallated carbon are called *carbenoids* [2]. At the beginning first their decompositions and reactions resembling those of carbenes were studied [3]. The three modes of reaction of lithium halocarbenoids, (with bromine as the halogen atom) which are generally stable only in the temperature range below —80 °C, are presented in *Scheme 1*. The geminal position of a positive (Li) and a negative (halogen) leaving group conveys both donor and acceptor properties to the carbon atom. According to *Scheme 1*

a) Warming of carbenoid solution in the presence of olefins leads to cyclopropanes with retention of configuration of the $C_2$-component.
b) In reactions with electrophiles, the configuration at the carbenoid centre is retained.
c) Strong nucleophiles can substitute the halide with formation of another organolithium compound.

The reactivity of carbenoids is thus equally characterized by electrophilic [(a) and (c) in *Scheme 1*] and by nucleophilic behaviour (b). The variations of the groups R (H, alkyl, cycloalkyl, cyclopropyl, vinyl, heteroatoms) furnish a remarkable multiplicity.

Since the preparative applications of lithium halocarbenoids have strongly increased in recent years, it is the purpose of this review to demonstrate the synthetic significance of these intermediates.

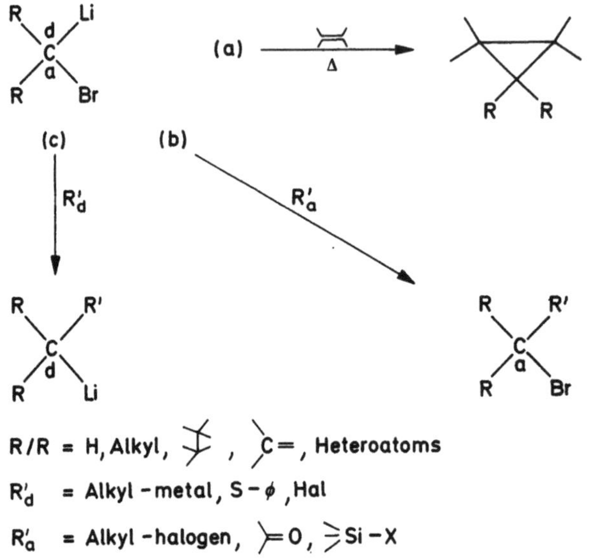


*Scheme 1*: Acceptor and donor reactivity of lithium halocarbenoids

# B. Preparative Reactions of Lithium Halocarbenoids

## B.I Carbon Atom Insertions

The carbenoid reactions leading to cyclopropanes *inter*molecularly have been reviewed extensively [4], so that we can restrict our discussion to some interesting *intra*molecular reactions, the so-called *carbon atom insertions*[5] or *naked carbon-atom insertions* [6] (for instance Eq. (1)):

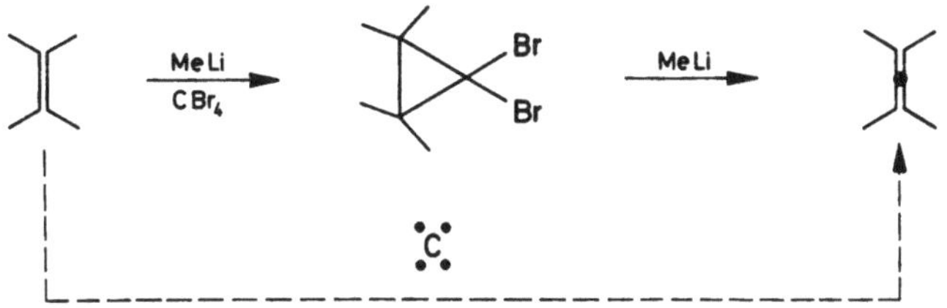

(1)

Formally, this synthesis of allenes [7] involves insertion of a carbon atom between the atoms of a double bond. If a too highly strained allene were formed, insertion into adjacent CH-bonds would occur instead [8]. Two recent examples of the well-known reaction are described by Eqs. (2) [5] and (3) [9]. As shown by *Cory* et al. the conversions can be carried out without isolation of intermediates if a twofold excess of the reagent $CBr_4/CH_3Li$ is employed to account for losses resulting from the decomposition of the intermediate $LiCBr_3$. Two further remarkable applications, in which bicyclopentane rather than bicyclobutane is formed, have been reported by *M. S. Baird* and his collaborators (Eqs. (4) [10] and (5) [11]).

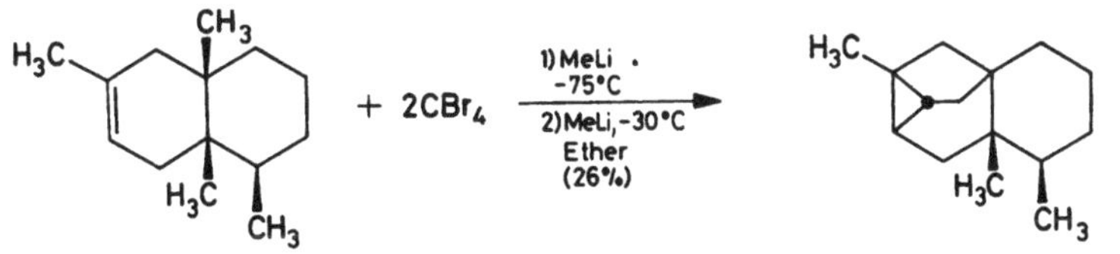

(2)

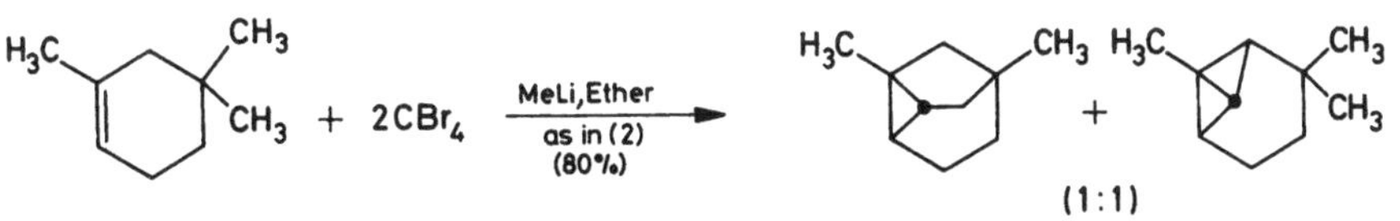

(3)

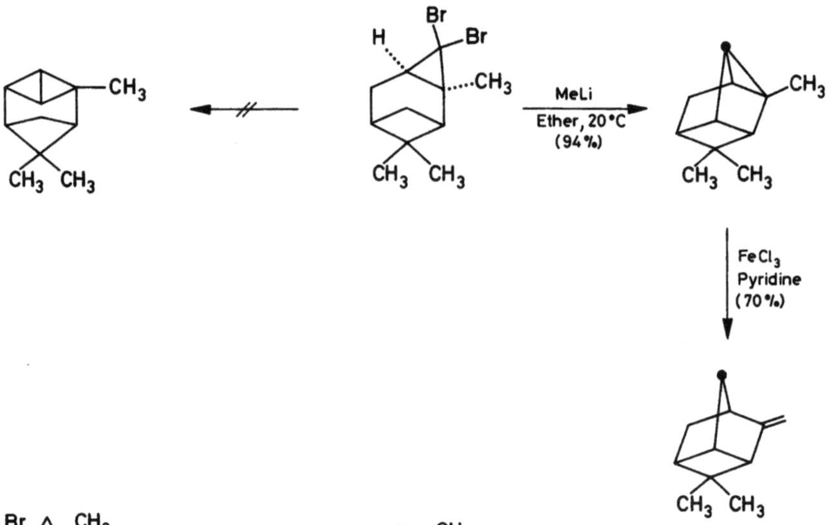


(4)

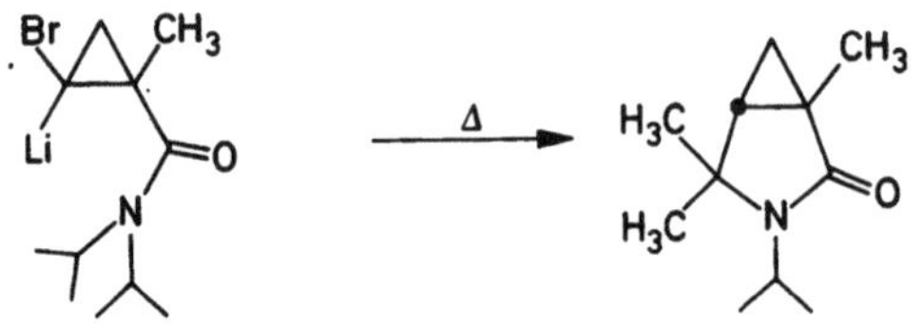


(5)

Finally, *Paquette* et al. have described some surprisingly selective carbon insertions in the propellane series (Eqs. (6)–(8)[12]).

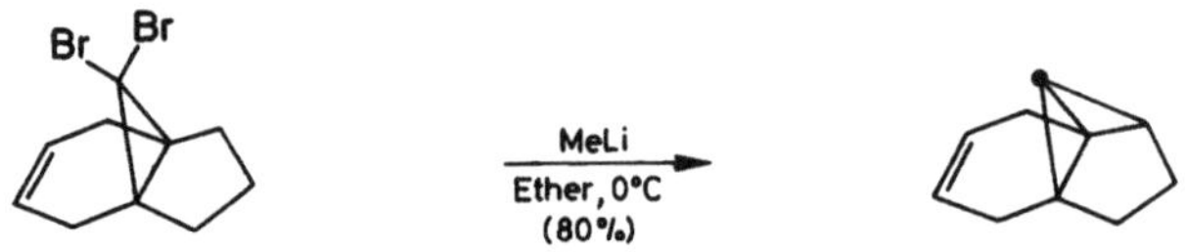


(6)

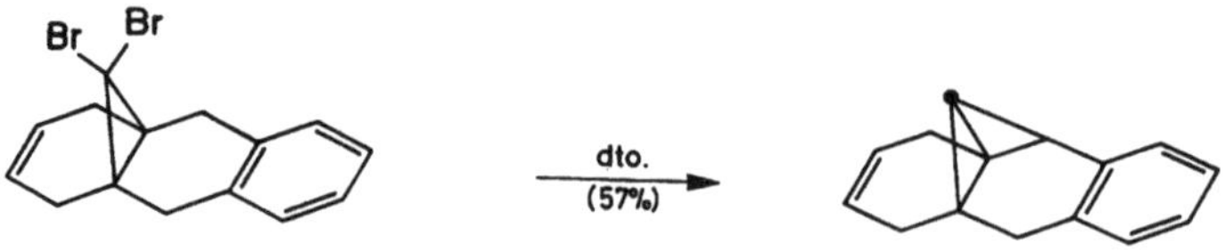


(7)

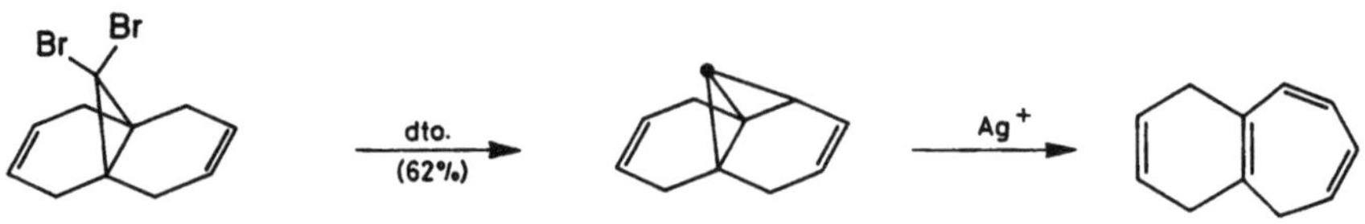


(8)

## B.II Reactions of Lithium Carbenoids with Electrophiles

*Köbrich* and *Trapp* were the first to prove the existence of halocarbenoids in solution by trapping experiments with electrophiles in 1963 [13]. Due to the existence of halocarbenoids only at low temperatures, it took a decade before this class of α-heterosubstituted organolithium compounds was recognized as synthetically useful reagents. Since the chemical and physical properties and hence the applicability of these species depend very strongly on the carbon skeleton and the other substituents at or near the carbenoid carbon, the following discussion is organized according to the groups bearing the geminal lithium and halogen atoms.

### B.II.1 Alkylidene Lithium Halocarbenoids

Formally, lithium halocarbenoids are obtained by replacing α-hydrogen atoms in alkyllithium compounds by halogenes. Thus, we can derive from methyllithium mono-halo-, dihalo- and trihalolithiomethanes

$H_2CHal\,Li$ $\qquad$ $HCHal_2Li$ $\qquad$ $CHal_3Li$

In addition, mono- and dialkyl-halocarbenoids

have recently obtained some attention of preparative chemists although they are extremely unstable (decomposition > —110 °C). An *in situ* application of a dialkyl substituted bromo carbenoid is *Cainelli's* preparation of oxiranes (see, for example, Eq. (9) [14]) while *Villieras* and his group have used solutions of chloro and bromo monoalkyl carbenoids for the preparation of α-haloketones and -aldehydes, according to Eq. (10) [15].

(9)

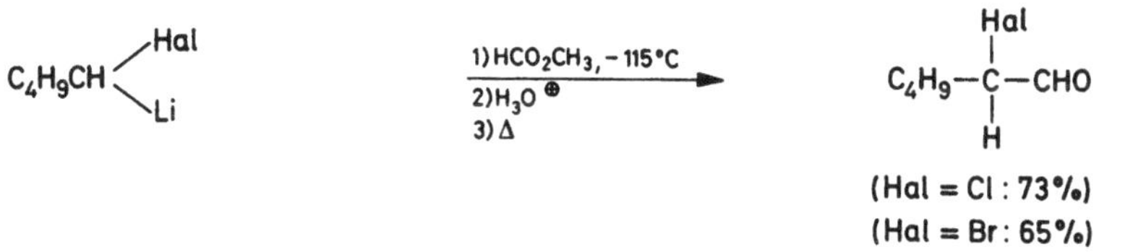


(10)

The types of reactions exemplified by Eqs. (9) and (10) are characteristic of all types of lithium halocarbenoids (see also below):

a) simple additions to carbonyl groups resulting in hydroxyalkylation or acylation;

b) formation of epoxides from the primary adducts to aldehydes or ketones, i.e. a formal carbene addition ("methylenation") to the carbonyl group.

Although highly reactive, *trihalo*methyllithium derivatives appear to be of limited preparative scope; the products formed on addition to various electrophiles [3,16] (Eq. (11)) can serve as precursors of the valuable α,α-*dihalo*-lithioalkanes, which are also available by H/Li-exchange in 1,1-dihaloalkanes using lithium amides. Although *Köbrich*'s early work [17] describes the

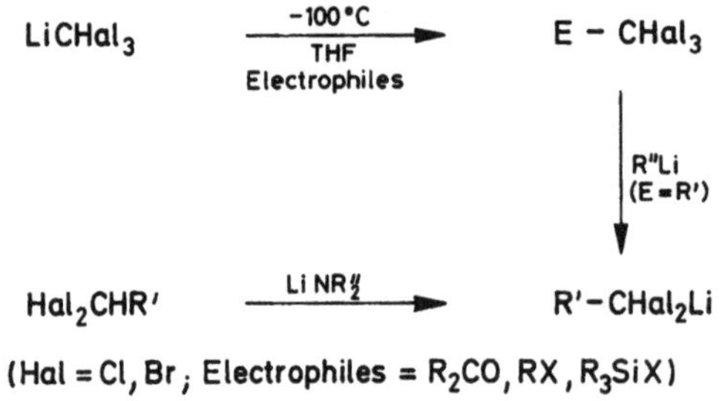


(11)

principal reactions of this class of carbenoids, the groups of *J. F. Normant* [16,18] in France and of *H. Nozaki* [18] in Japan elaborated their applications in synthesis. The formation of simple alkyl, hydroxyalkyl, and acyl derivatives is shown in Eq. (12). Most versatile are the additions of α,α-dihalo-lithioalkanes to carbonyl

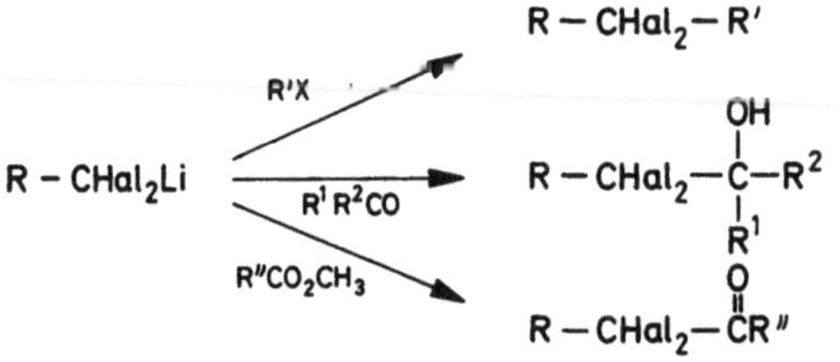


(12)

groups of aldehydes or ketones because the primary adducts are amenable to subsequent transformations:

a) An overall carbonylmethylenation to a halo epoxide and a carbonyl transposition are illustrated by Eq. (13) [16].

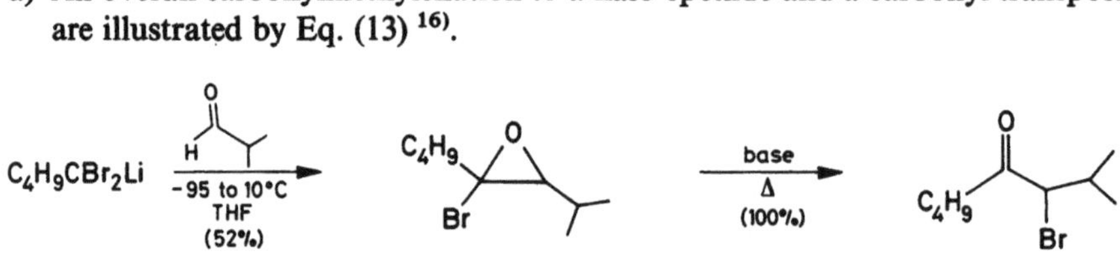


(13)

b) Overall carbonyl olefinations, i.e. *Wittig*-type reactions, using dichloro- or dibromomethyllithium are described in several papers [19,20]. As is evident from Eq. (14-1) etherification of the primarily formed lithium alkoxides and halogen/lithium exchange lead to *carbenoids* (lithioalkanes with an α-leaving group) which at the same time are *olefinoids* (lithioalkanes with a β-leaving group). These are also produced by methoxyalkylation of an alkyl dihalocarbenoid and subsequent deprotonation with a lithium amide (route (14-2)). Upon warming from —110 to —60 °C, the *olefinoids* eliminate lithium alkoxides to give geminal dihaloolefins [20].

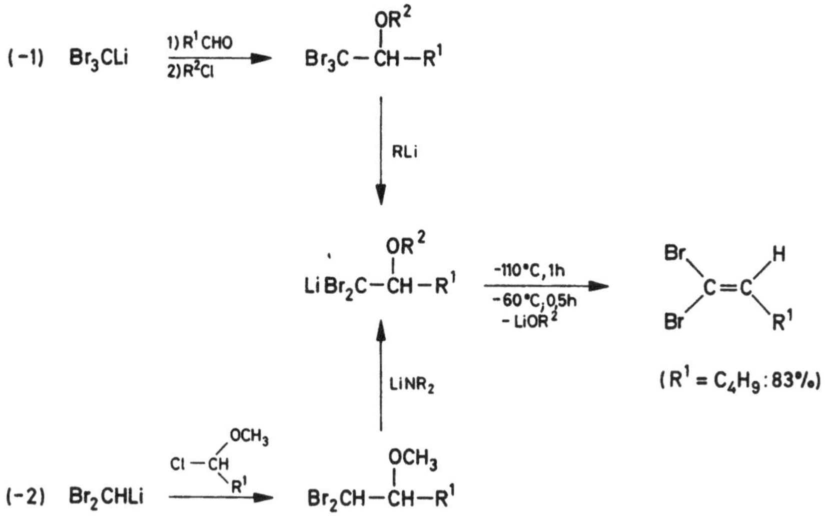


(14)

c) Halo-homologation (with $Hal_3CLi$) or homologation (with $Hal_2CHLi$) is the overall result if halogen/lithium or H/lithium exchange is carried out at the stage of the lithium alkoxides, the primary carbonyl adducts. Since $Li_2O$ is a very poor leaving group, α-LiHal elimination instead of β-elimination takes place with (simultanous?) rearrangement to a lithium enolate (Eq. (15)). According to

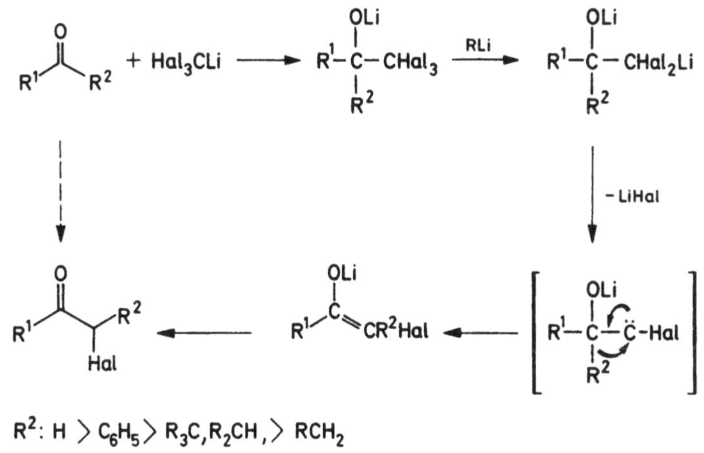


(15)

*Normant* et al. [19], the migration power decreases in the order given underneath Eq. (15), and is reminiscent of that in other base-induced sextet rearrangements (benzilic acid, *Favorsky*-type).

The reaction was used by *Nozaki* et al. [21] for developing a new method of ring enlargement by one carbon atom (Eqs. (16)–(18)). *Normant*'s homologative acetylene synthesis (Eq. (19-1)) and *Kowalski*'s homologation of carboxylic esters (Eq. (19-2)) are based on the same principle [22] (cf. *Fritsch-Buttenberg-Wiechel* rearrangement [3]).

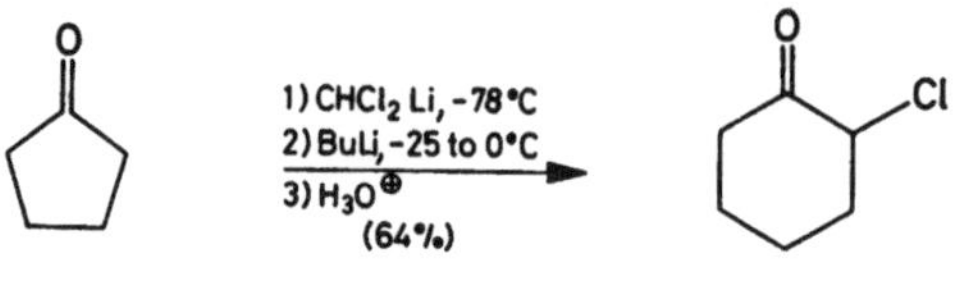


(16)

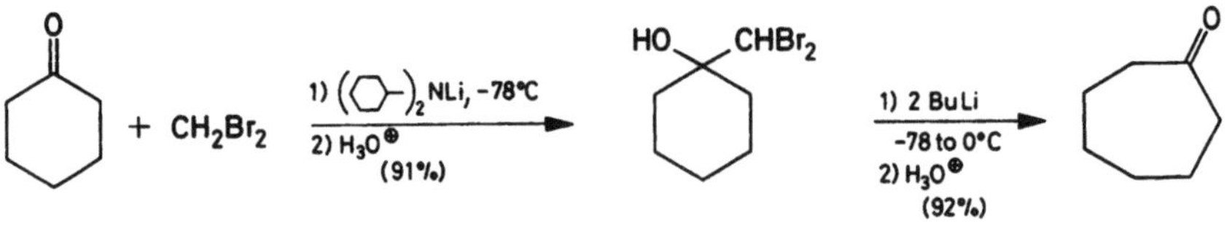


(17)

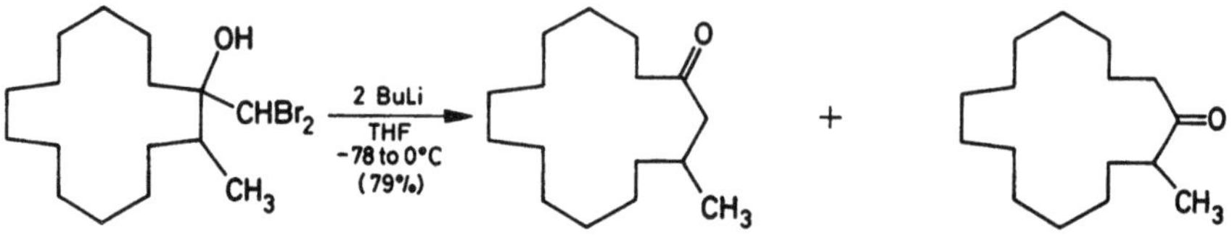


(18)

97 3

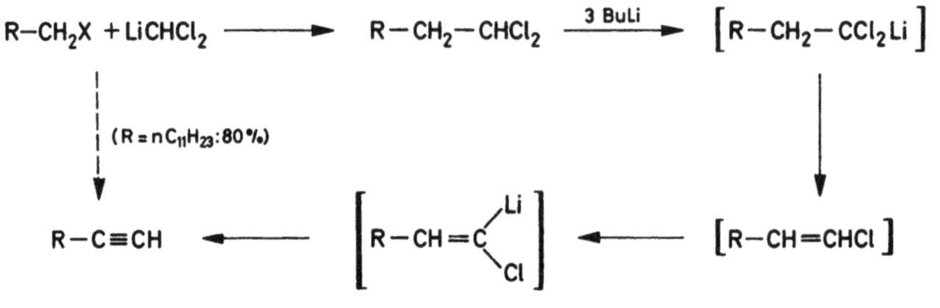


(19-1)

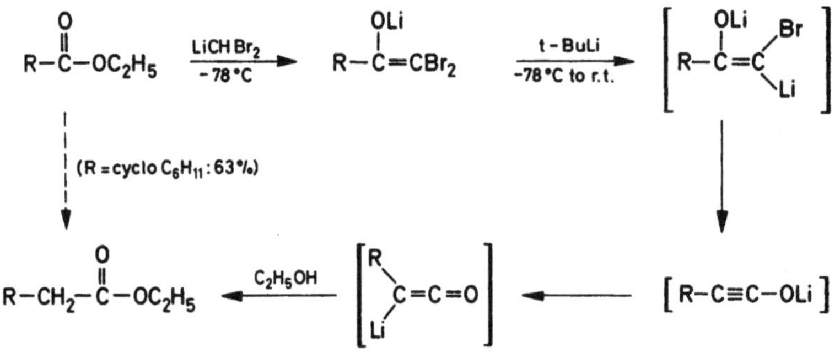


(19-2)

Finally treatment of α,α-dibromomethyl alkohols with zinc and glacial acetic acid causes β-elimination of HOBr yielding E- and Z-bromo-olefins, as shown by *D. R. Williams* et al. for a variety of compounds [23]:

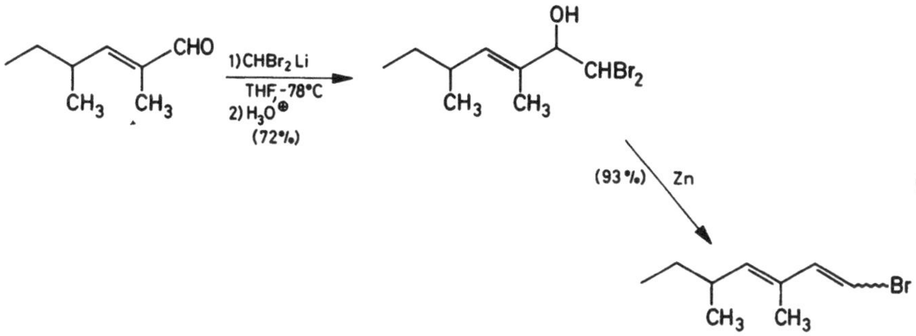


(20)

This chapter should not end without mentioning that the products of alkoxycarbonylation of lithium halocarbenoids can be converted to enolates reacting with various electrophiles to give for instance the α-chloro and α,α-dichloroesters as shown by Eqs. (21) and (22) [24]. Although we are probably dealing with chloro-substituted lithium enolates and not with α-chloroorganolithium compounds, enolates carrying one chlorine appear to be more stable than their dichloro analogs [24].

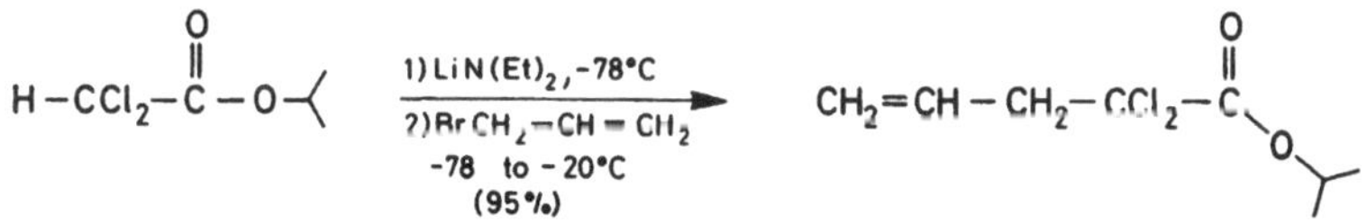


(21)

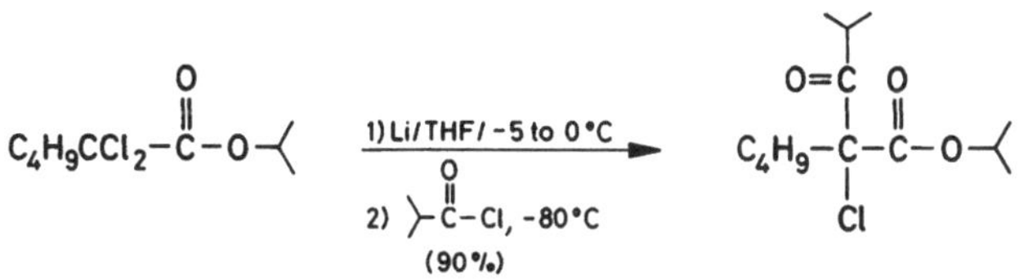


(22)

B.II.2 α-Heteroatom-Substituted Lithium Halocarbenoids

An extension of the synthetic applicability of lithium halomethanes is achieved by the simultanous presence of another main group heteroatom at the same carbon. Thus, if one of the chlorine atoms of dichloromethyllithium is replaced by a sulfonylamin group, the following products are obtained by reaction with electrophiles (Eq. (23)) [25]. The substituted carbenoid can be converted to normal carbonyl adducts as well as to olefins and cyclopropanes.

Lithiohalomethyl phenyl sulfoxides have widely been applied. Thus, *V. Reu-*

*trakul* et al. used them for the homologisation of aldehydes to α-haloketones [26] and for the synthesis of α,β-unsaturated aldehydes from ketones [27] (Eqs. (24) and (25)). It should be noted that, formally, LiSOR rather than LiHal is eliminated in Eq. (24).

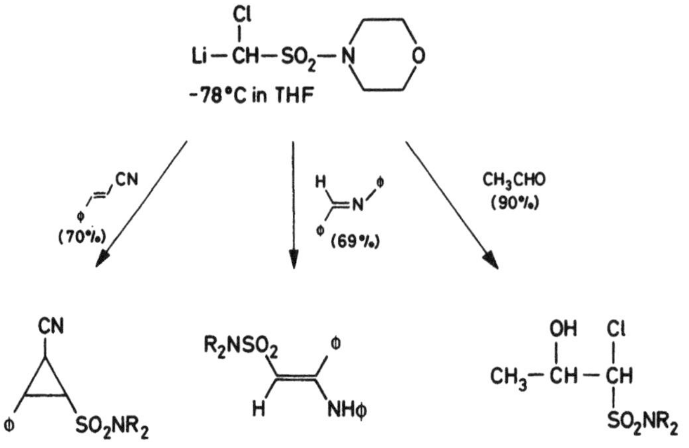

(23)

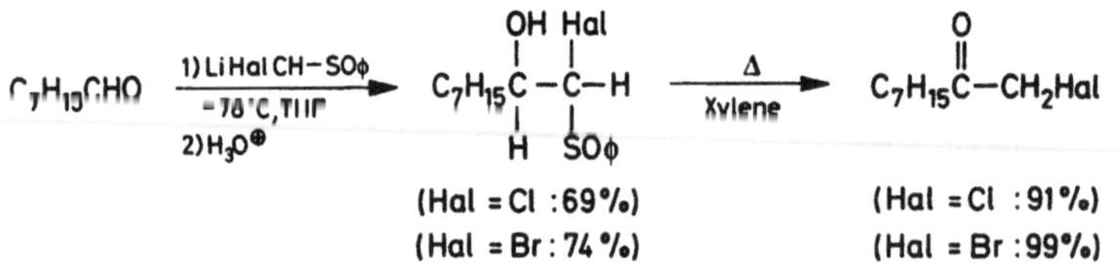

(24)

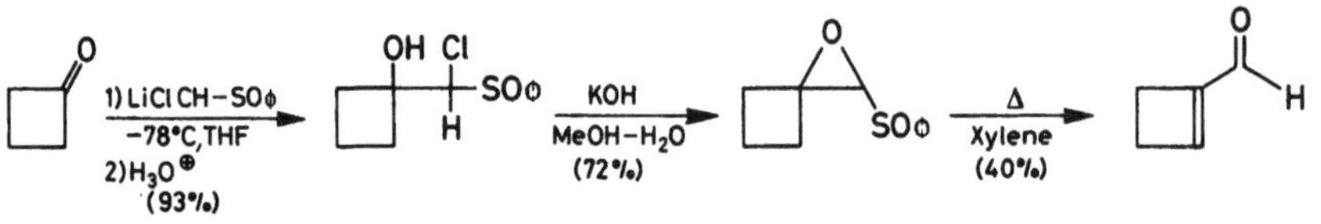

(25)

Another modification of the reaction shown above is found in the work of *T. Durst* et al. [28] and *D. F. Taber* and *B. P. Gunn* [29] who obtained α-bromocarbonyl derivatives, γ-carbonylsulfoxides and sulfones as well as α,β-unsaturated ketones, as outlined in Eq. (26).

In a sequence of steps, including a silyl *Pummerer*-rearrangement, *Wemple* et al. used chloromethyl phenyl sulfoxide as a phenyl thiocarbonyl-$d^1$-reagent, as shown in Eq. (27) [30].

The application of α-silyl-α-chloroalkyllithium reagents (*Magnus* et al., Eq. (28)), is yet another reductive nucleophilic acylation of carbonyl groups [31]. Reagents of this type were actually first described 15 years ago by *Köbrich* and his group [3]. The use of $Me_3SiCH(OCH_3)Li$ for the same transformation may be advantageous [32].

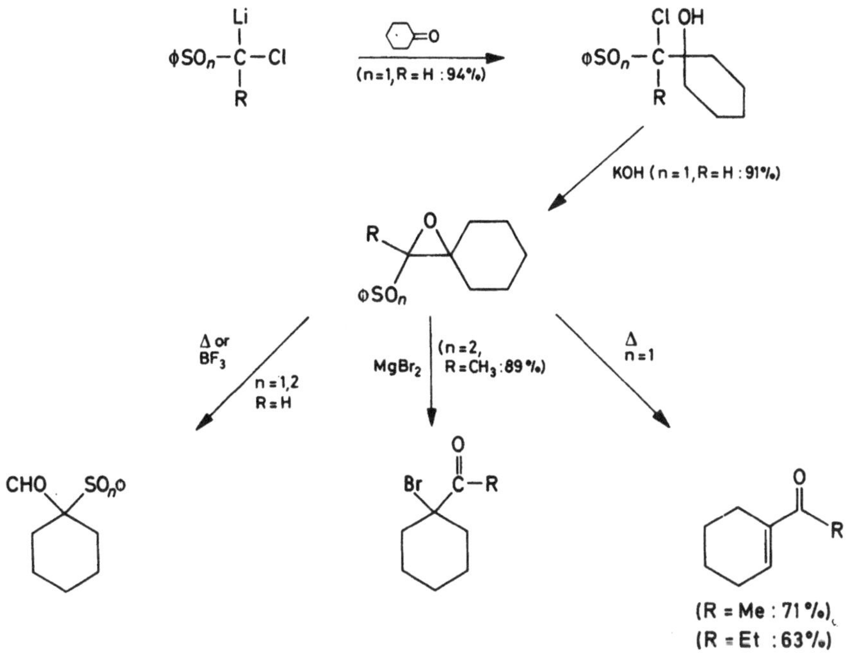

(n=1, R=H : 94%)
KOH (n=1, R=H : 91%)
Δ or BF3
n=1,2
R=H
MgBr2
(n=2, R=CH3 : 89%)
Δ
n=1
(R = Me : 71%)
(R = Et : 63%)


(26)

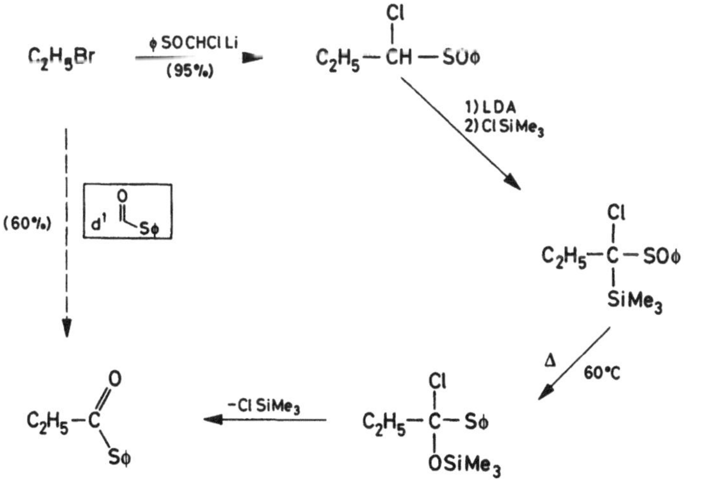

C2H5Br
ϕSOCHClLi
(95%)
1) LDA
2) ClSiMe3
(60%)
Δ
60°C
-ClSiMe3


(27)

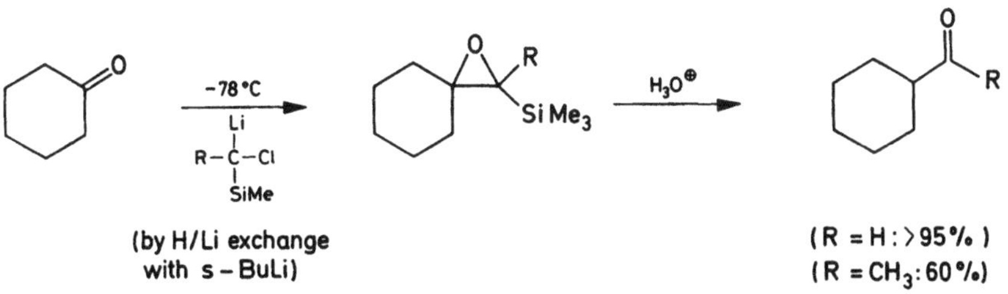

-78°C
H3O⊕
(by H/Li exchange with s-BuLi)
(R = H : >95%)
(R = CH3 : 60%)


(28)

With borates the silyl substituted chloro carbenoid yields the insertion product (Eq. (29) [33]).

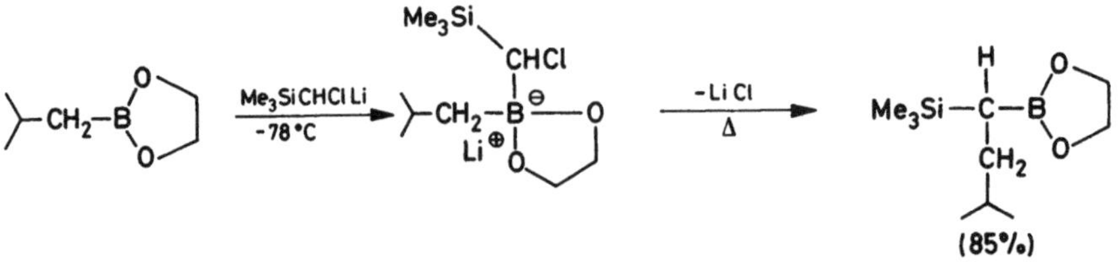
(29)

In reaction (28) γ-elimination of LiCl with the formation of an epoxide is obviously faster than β-elimination of $Me_3SiOLi$ (*Peterson*-olefination). *Seyferth* and his collaborators [34] found that this reaction takes place when bis(trimethylsilyl)bromomethyllithium is reacted with carbonyl compounds, albeit not stereoselectively (Eq. (30)). In contrast, thermal dehydrohalogenation of 1,1-dichloro-1-phenyldimethylsilyl-alkanes furnishes Z-olefins only (Eq. (31)) according to the results of *Larson* et al. [35]:

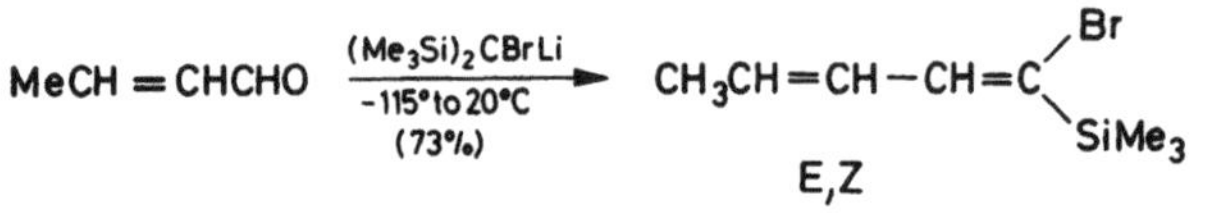
(30)

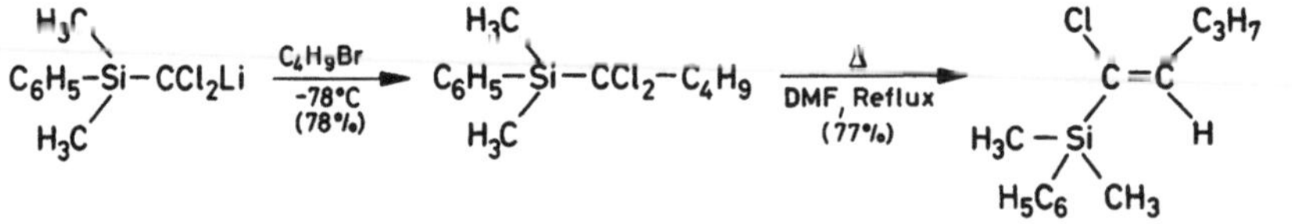
(31)

(32)

Higher yields in reactions of trimethylsilyl-bromomethyllithium are obtained by using THF/ether mixtures and sec.-butyllithium for the Br/Li-exchange at −120 °C, as outlined in Eq. (32) [36].

Finally, the synthetic importance of phosphorous-substituted lithium halocarbenoids has been demonstrated by making extensive use of the reactions of diethyl α-monohalo- and α,α-dihalolithioalkylphosphonates with electrophiles.

Diethyl dichlorolithiomethanephosphonate is readily prepared by Hal/Li exchange from diethyl trichloromethanephosphonate at temperatures below −85 °C and can be reacted in the presence of one equivalent of HMPT with a variety of alkyl halogenids [37].

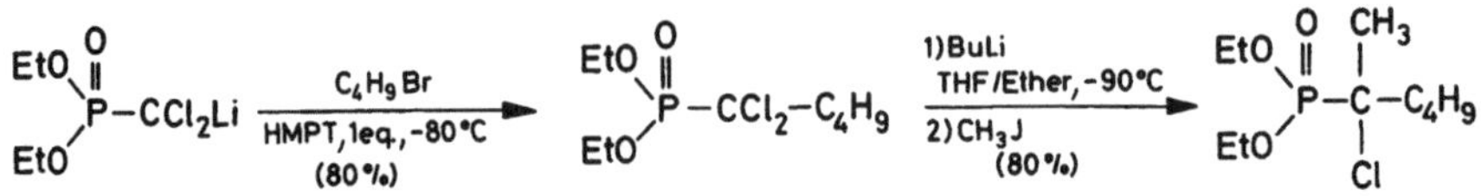
(33)

As seen from Eq. (33), the reaction sequence can be performed one more time with the monoalkylated products to yield the bisalkylated phosphonates [37].

On the other hand, bisalkylation of diethyl 1-chloro-1-lithioalkanephosphonates can also be achieved by copper(I)-catalysed replacement of chlorine by carbon-nucleophiles with subsequent alkylation of the resulting lithiocarbanion [38] (Eq. (34)).

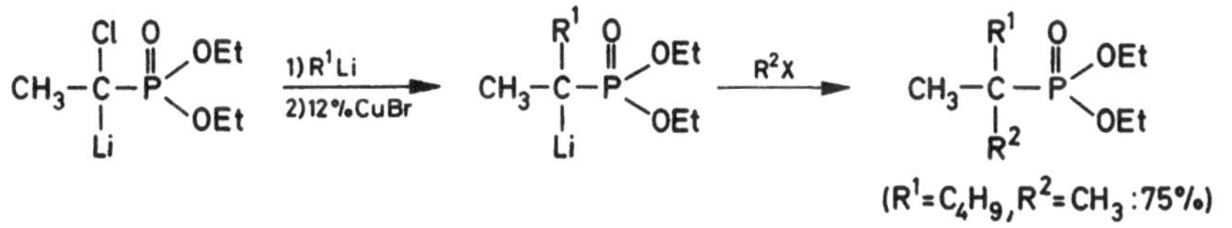


(34)

A very interesting synthetic application of diethyl dichlorolithiummethylphosphonate is described by *Villieras* et al. In a one-step procedure the carbenoid can subsequently be reacted with ethyl carbonochloridate or a carboxylic acid chloride and an aldehyde followed by a *Wittig-Horner* elimination on warming the reaction mixture. In this way, it is possible to obtain α-chloro-α,β-unsaturated ketones and esters in good yields [39]:

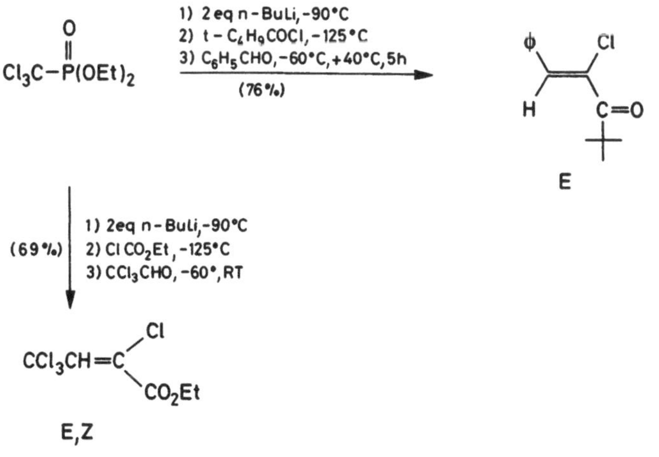


(35)

The reaction proceeds via an enolate produced as intermediate as shown by its protonation [39] (Eq. (36)).

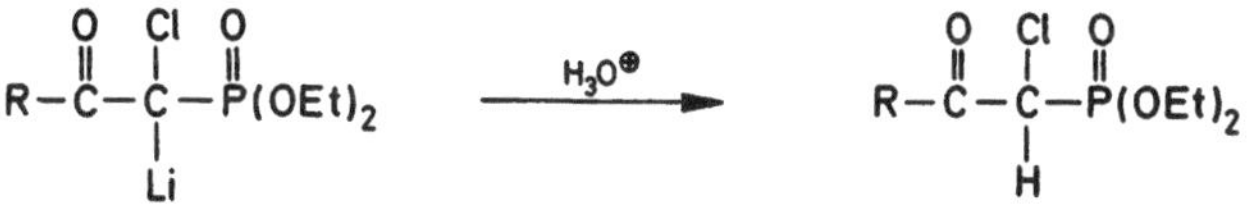


(36)

The same authors have reported the reactions of 1-chloro-1-lithioalkanephosphonates with carbonyl compounds yielding either the respective chlorohydrins or diethyl 1,2-epoxyalkanephosphonates, depending on the reaction conditions and not the expected chloroolefins (Eq. (37)) [40].

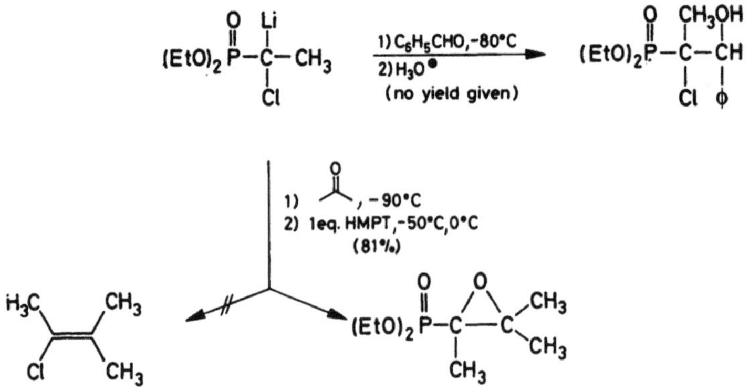


(37)

A more convenient procedure without the use of HMPT leading to diethyl 1,2-epoxyalkanephosphonates has been reported by *P. Savignac and P. Coutrot*[41] who also showed that a *Wittig-Horner* olefination takes place when diethyl *dihalo*-methyllithiumphosphonates are reacted with carbonyl compounds, as shown by Eq. (38)[42].

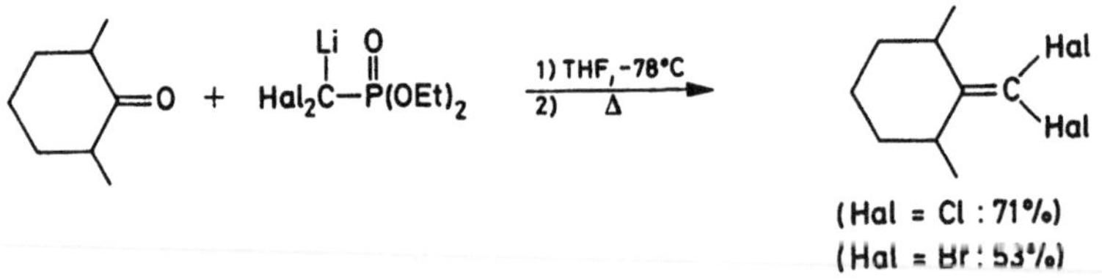


(38)

B.II.3 Cyclopropylidene Lithium Halocarbenoids

This class of carbenoids was investigated by *Köbrich* and his group[43] long before its synthetic usefulness was recognized. They studied most extensively the 7-chloro-7-lithionorcarane.

1-Lithio-1-halocyclopropanes react with electrophiles as expected of organolithium compounds. Thus, *Seyferth* et al.[44] prepared a series of silicon, tin, and lead derivatives according to Eq. (39). They also studied the stereochemistry (endo/exo-

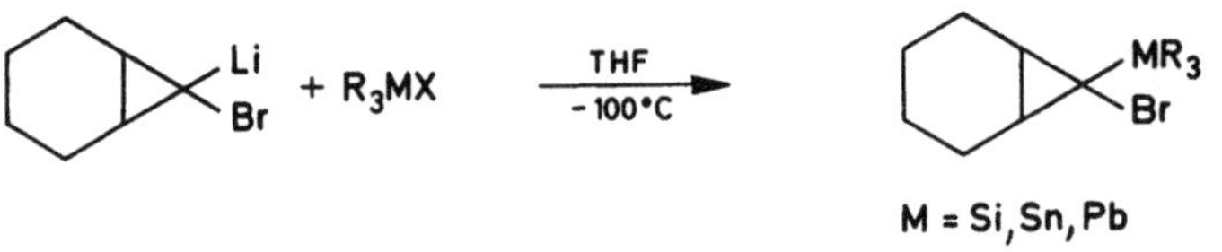


(39)

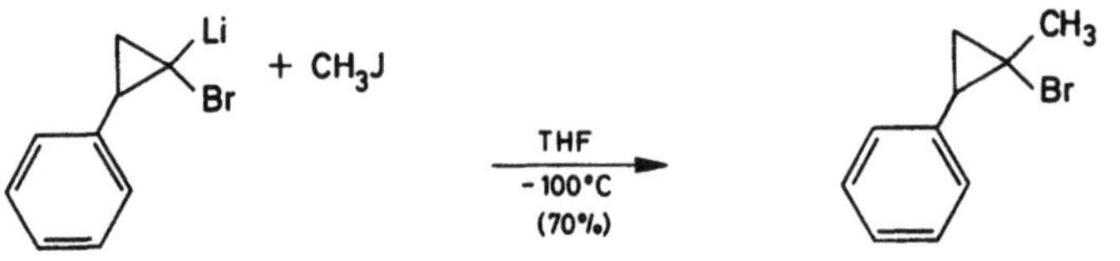


(40)

isomerism) of the reaction products. The alkylation (electrophilic substitution) outlined in Eq. (40) was described by *Nozaki* et al. [45] who also discovered that excess of organometallic reagents can lead to nucleophilic substitution at the carbenoid center [46] (see examples in Eqs. (41) and (42)); these reactions proceed with a remarkably high stereoselectivity [47].

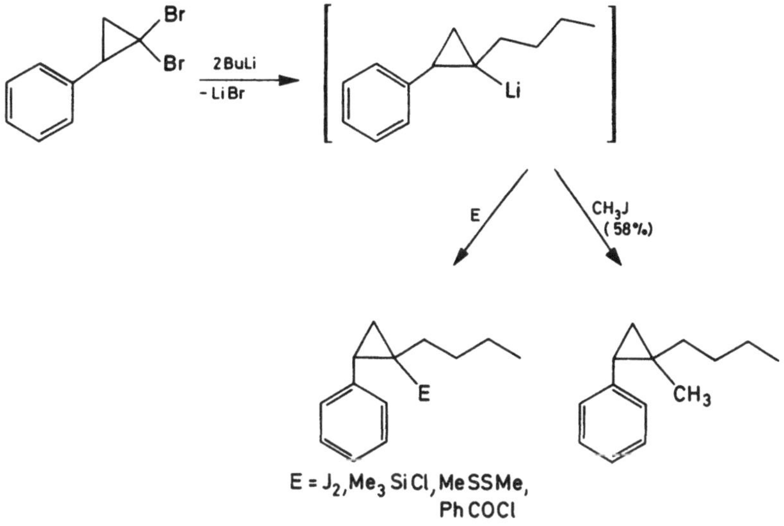


(41)

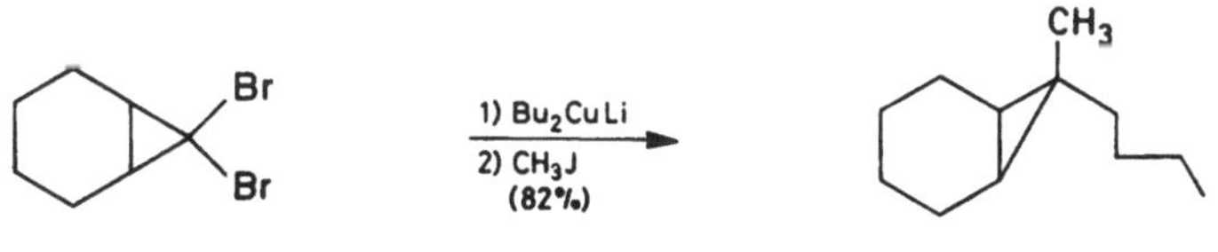


(42)

In recent years, *Seebach* et al. have elaborated synthetic applications of α-halo-lithiocyclopropanes, for instance, insertion of a CO-group between the two vinylic carbon atoms of an olefin (Eq. (43) [48]), annulations of four-membered rings and of lactones (Eq. (44) [49]), preparation of cyclopropylketones from olefins (Eq. (45) [49]), and, by using the same cyclopropylbromohydrins, bromovinylations [50] and synthesis of methylenecyclopropanes from olefins, bromoform and aldehydes (Eq. (46) [51]).

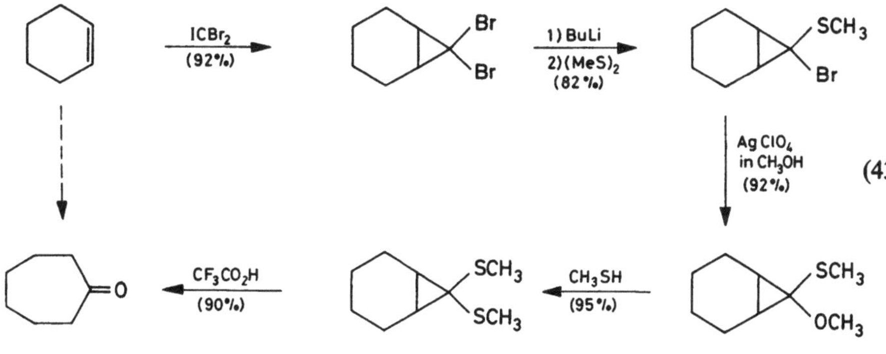


(43)

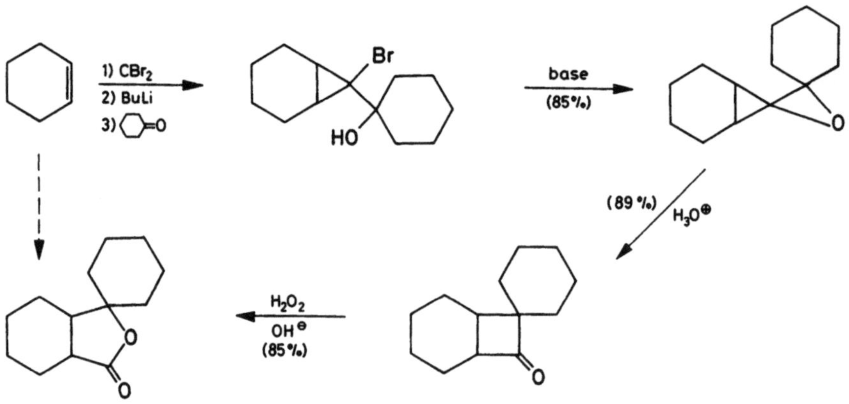
1) CBr2
2) BuLi
3)
Br
HO
base
(85%)
O
(89 %)
H3O⊕
H2O2
OH⊖
(85%)
O
O
O

(44)

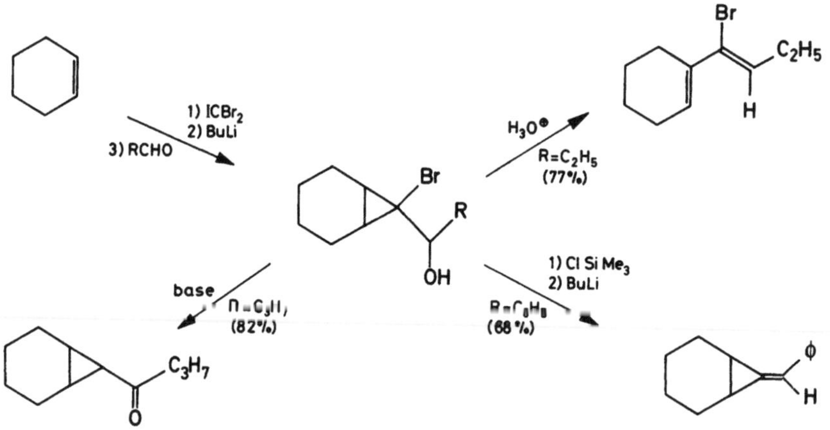
1) ICBr2
2) BuLi
3) RCHO
Br
R
OH
H3O⊕
R=C2H5
(77%)
Br
C2H5
H
base
(82%)
C3H7
O
1) Cl Si Me3
2) BuLi
(68 %)
ϕ
H

(45)

(46)

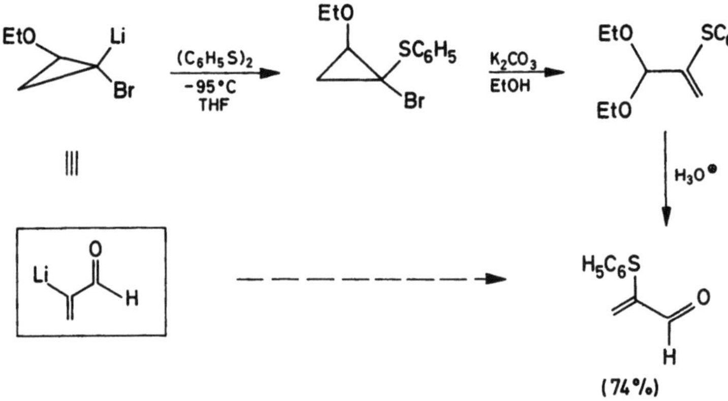
EtO
Li
Br
(C6H5S)2
-95°C
THF
EtO
SC6H5
Br
K2CO3
EtOH
EtO
SC6H5
EtO
H3O⊕
O
Li
H
H5C6S
O
H
(74%)

(47)

The preparation of α-thiolated acrolein [52] and the methylene lactone annulation [53] by *Hiyama* et al. are two further examples of the applications of α-halocyclopropyllithium derivatives which have to be handled at temperatures well below dry ice temperature (Eqs. (47) and (48)).

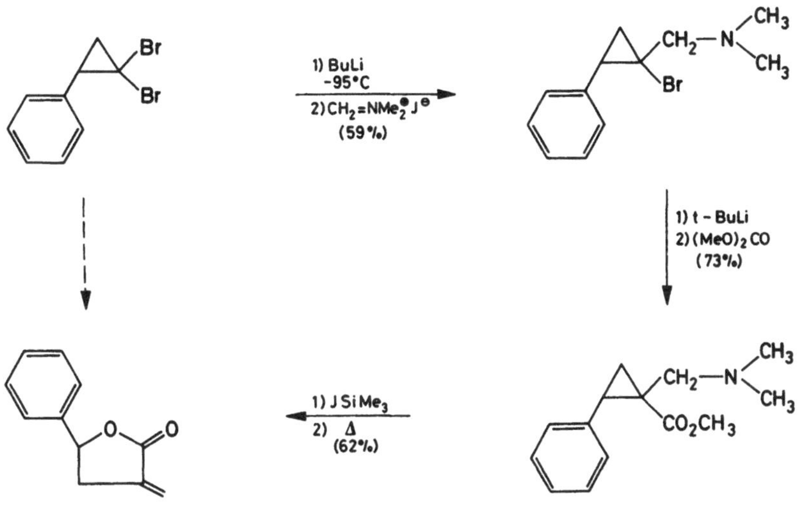


(48)

B.II.4 Vinylidene and Allylidene Lithium Halocarbenoids

Vinylidene lithium halocarbenoids may be generated by halogen/lithium exchange from the corresponding dihaloolefins. In addition, there is another route to these intermediates involving metallation of monohaloolefins, since the acidity of the vinylic proton is sufficiently high (Eq. (49) [3]).

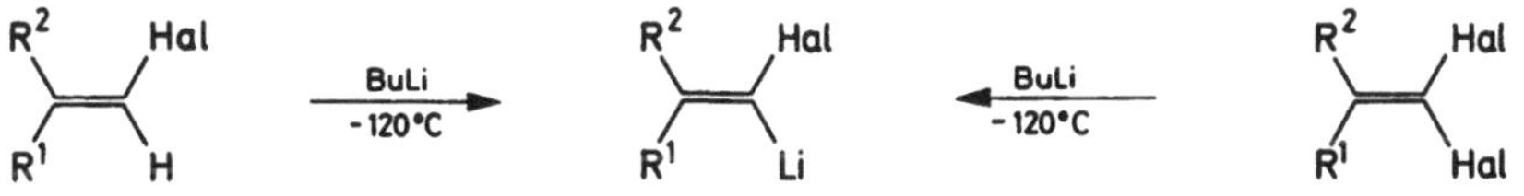


(49)

$R^1, R^2$ = Alkyl, Aryl, Heteroatoms

Several synthetic applications of these intermediates have been reported by *J. F. Normant* and coworkers. For instance, trifluorovinyllithium reacts with carbonyl compounds in the usual manner to give carbinols. When treated with concentrated sulfuric acid these primary adducts undergo allylic rearrangement to the corresponding acid fluorides which, in turn, can be converted to acids, esters or amides as outlined by reaction sequence (50) [54].

Instead to acid fluorides the reaction between trifluorovinyllithium and carbonyl compounds can lead to α-fluoro-α,β-unsaturated aldehydes and ketones by treating the intermediate lithium alkoholates either with a lithiumalkyl or lithium aluminium hydride, as the geminal fluoro atoms on the vinylic carbon in the alcoholates are susceptable to nucleophilic replacement by these reagents (reaction sequence (51)). Finally, rearrangement yields the α-fluoro-α,β-unsaturated aldehydes and ketones [55].

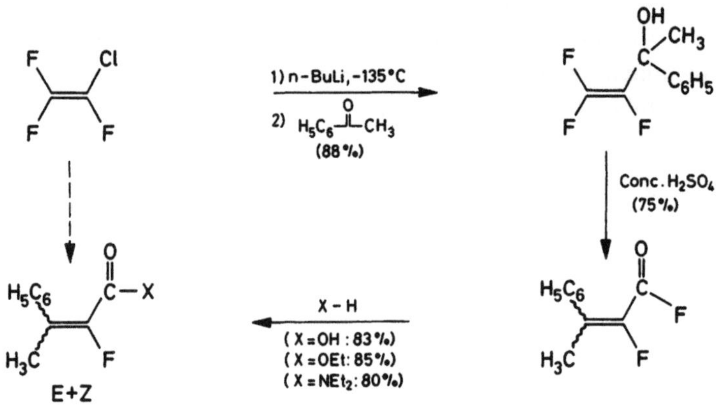
(50)

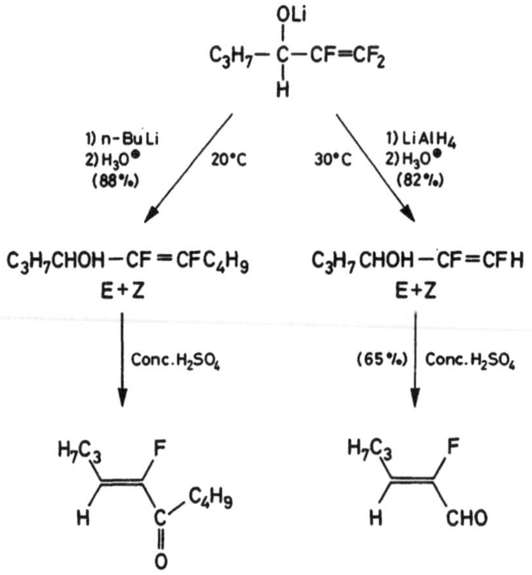
(51)

The same reaction sequence also proceeds when starting with 1-chloro-1-lithio-difluoroethylen generated from 1,1-dichlorodifluoroethylen and butyllithium, thus providing an access to α-chloro-α,β-unsaturated carbonyl compounds [56].

From the reaction of perfluoro-1-lithiopropene with several electrophiles (Eq. (52)) it was unambiguously shown by *Hahnfeld* and *Burton* that vinylidene lithium fluorocarbenoides react with retention of configuration, i.e. with stereochemical integrity, as known for their hydrocarbon analogs [57].

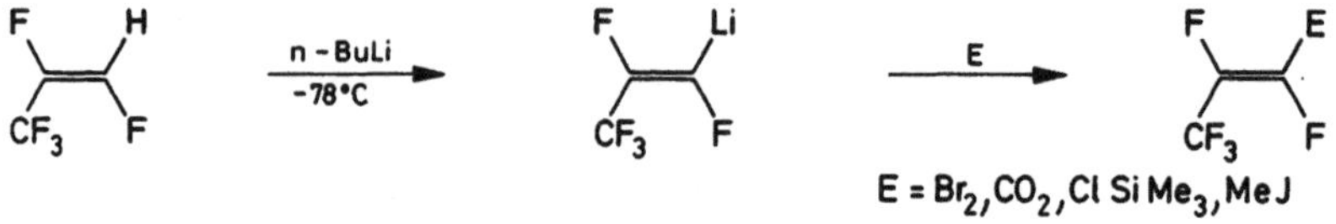
(52)

*Karrenbrock* and *Schaefer* prepared α-chloro-α,β-unsaturated esters by lithiation of diethyl 2,2-dichloro-1-ethoxyvinylphosphate and subsequent reaction with carbonyl compounds (Eq. (53)) [58].

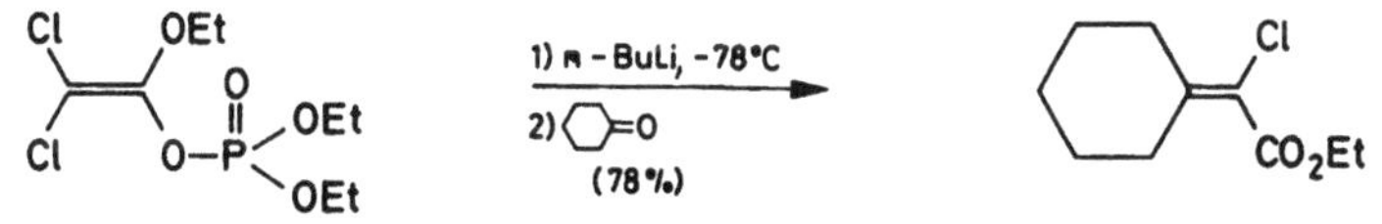


(53)

The phosphate reagent is obtained from trichloroacetate and triethyl phosphite. It can bring about the same transformation as diethyl 1-chloro-1-ethoxycarbonyl-1-lithiomethanephosphonate described by *Villieras* [39].

An interesting application of lithium halocarbenoids at temperatures above —80 °C involving substitution of halogen by alkylanions (cf. *Scheme* 1) and subsequent alkylation of the resulting carbanion has been reported by Duhamel [59] (reaction sequence (54)).

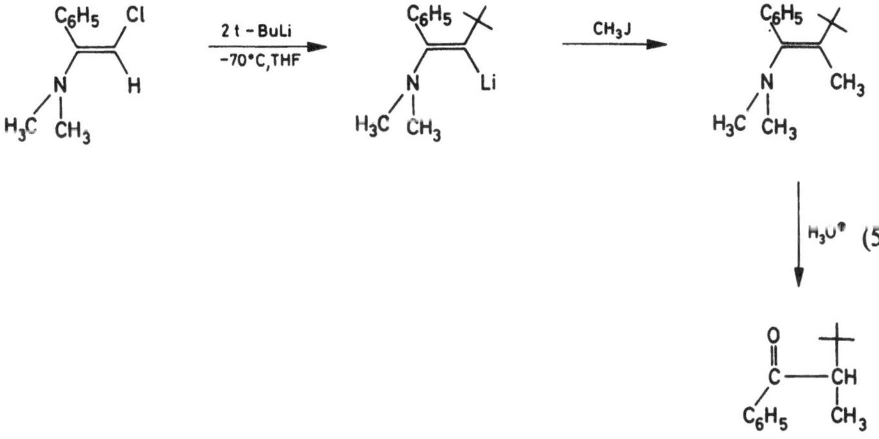


(54)

The overall reaction represents a one-pot double alkylation of α-chloroacetophenone.

An extension of the synthetic use of vinylidene lithium halocarbenoids is the method of *Zweifel* [60] who succesfully generated a vinylidenecarbenoid bearing a β-hydrogen atom, as outlined in Eq. (55). Usually, attempts to prepare these intermediates resulted in the elimination of hydrogen halide with the formation of the corresponding acetylene [3].

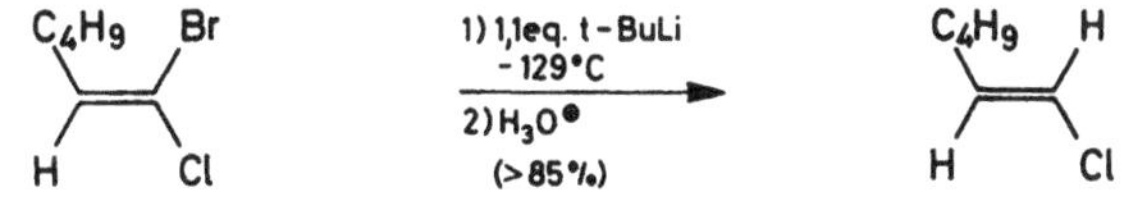


(55)

In contrast to their fluoro and chloro counterparts *bromo* lithium vinylidenecarbenoids have found little interest in organic synthesis. Nevertheless, the reaction of these carbenoids with trimethylchlorosilane was used by *Seyferth* et al. for the preparation of an α-bromovinylsilane (Eq. (56))[34].

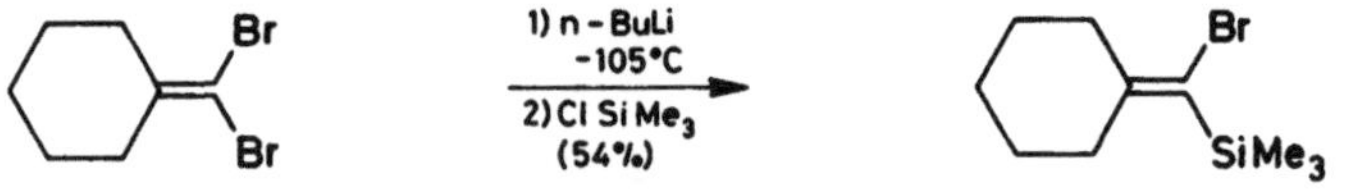


(56)

Reaction sequence (57) demonstrates that these intermediates have considerable synthetic potential, as shown by the synthesis of an allene with acetone, tetrabromomethane and benzaldehyde as starting materials[61].

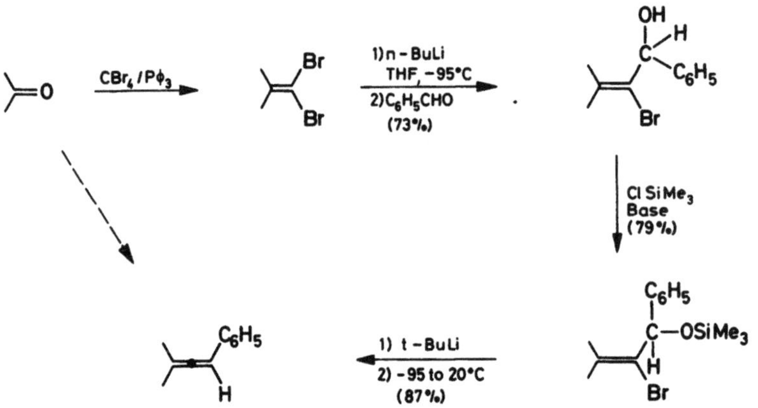


(57)

Somewhat similiar to the reactions of vinylidenecarbenoids are those of 1-lithio-1-haloallylidene carbenoids. According to *B. Mauzé*[62], 1-chloro-1-methallyllithium reacts with carbonyl compounds to yield E- and Z-vinyloxiranes, with aldimines vinylaziridines are formed (Eq. (58)).

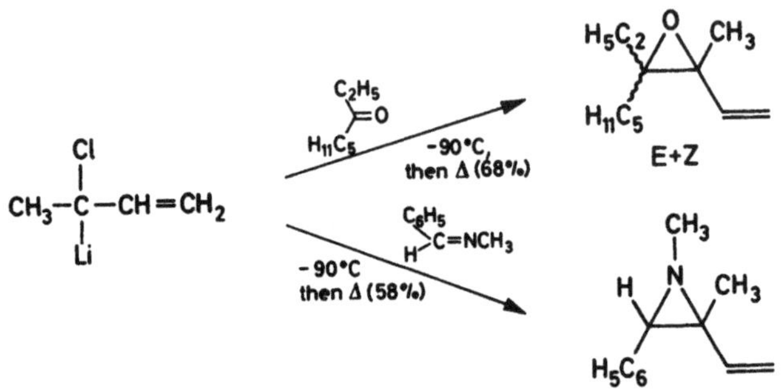


(58)

Likewise, 1,1-dichloroallyllithium preferably reacts C-1 with electrophiles[63] as is seen from the results of *Hiyama* et al. who used this reagent for cyclopentenone annulation[64] (Eq. (59)).

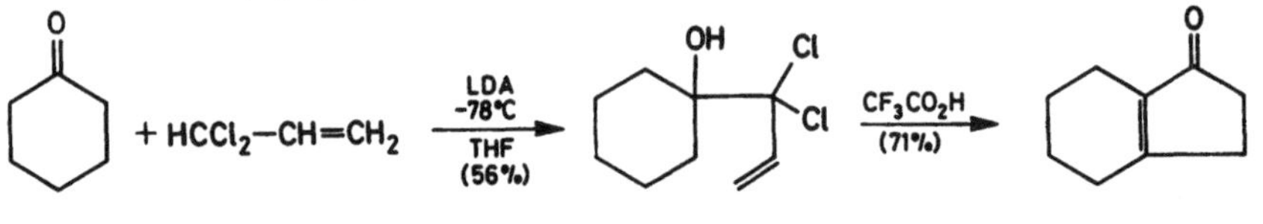


(59)

The rather unstable 1,1-difluoroallyllithium has been prepared by transmetallation from the corresponding tin derivative, according to *Seyferth* et al. [65]. These authors [66] have also reported a more efficient access to this carbenoid, and a possible applicability is achieved by Br/Li exchange reaction with 3,3-difluoroallyl bromide, as outlined by Eq. (60).

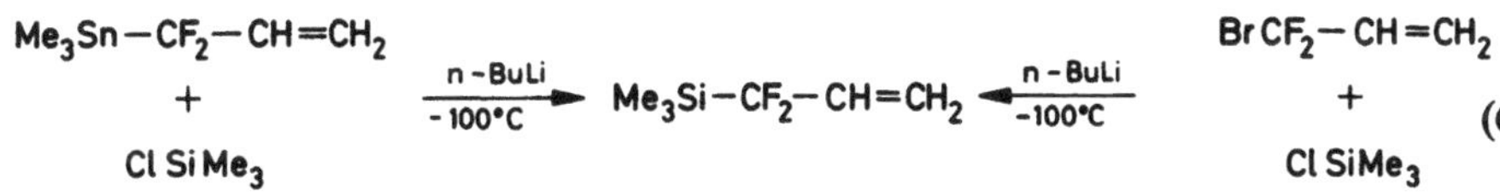


(60)

Valuable synthetic routes have been reported by *Duhamel*'s 3-bromo-3-lithium-1-azaallylidene-carbenoids [67], which, as seen from Eq. (61), are available from bromoaldimines.

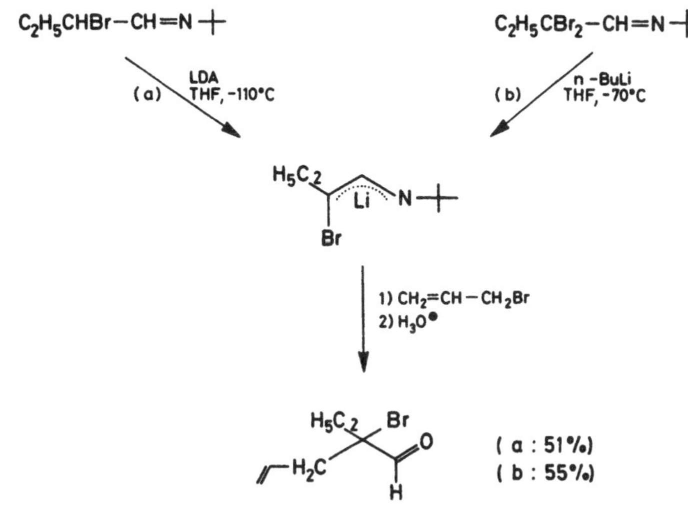


(61)

Finally, the results of *Wenkert* et al. show that bromoallylidenelithum undergoes rapid self-condensation when generated from lithium amide and allyl bromide [68] (Eq. (62)).

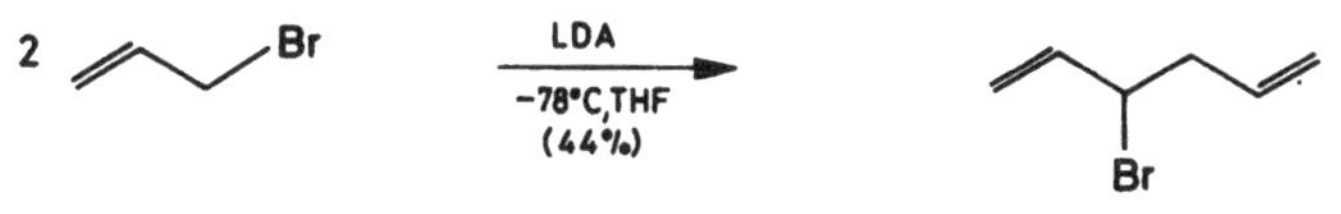


(62)

As is seen from the foregoing examples, 1-halolithiumallylidene carbenoids are preferably alkylated at the halogen-bearing carbon atom, i.e. they react as *$d^1$-reagents* [1].

## C. Conclusion

From the reactions of lithium halocarbenoids we have learned that they are reagents of high functionality. Depending on the type of carbenoid, up to four substituents can be replaced on the same carbon atom by consecutive electrophilic and nucleophilic reactions. A survey of these possibilities is given in *Scheme 2*.

| Reaction Type | Required Functionality | Polarity pattern | Carbenoid Type | Example (Reference) |
|---|---|---|---|---|
| Hydroxy-alkylation | 1 | ⊕<br>⊖ C—<br>\| | 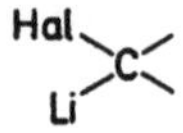 | (49) 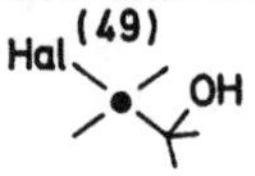 |
| Carbonyl methylen-ation | 2 | ⊕<br>⊖ C—<br>\| | 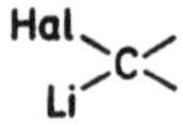 | (16) 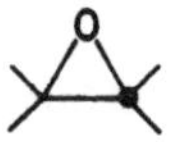 |
| Haloole-fination | 2 | ⊖<br>⊖ C ⊕<br>⊕ | *) 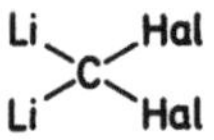 | (20) 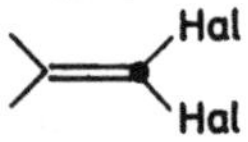 |
| Nucleophilic Acylation | 3 | ⊕<br>⊖ C ⊕<br>\| | 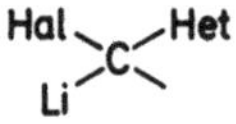 | (31) 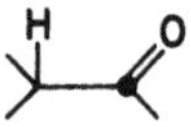 |
| Naked Carbon Insertion | 4 | ⊕<br>⊖ C ⊖<br>⊕ | *) 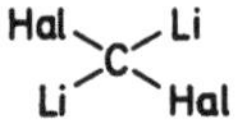 | (10)  |

*Scheme 2: Synthetic potential of lithium halocarbenoids*

* The schematically shown double lithiated centers are generated not at once under reaction conditions, i.e. the reaction proceeds via two monolithiated intermediates.

So in the last decade the preparative aspects of lithium halocarbenoids have been well established.

Some characteristics of their chemistry are still not well enough understood, as for example the nucleophilic substitution on the carbenoid center and its possible application in synthesis. Another still unsolved problem is the question of the structure of halo carbenoids. More insight into this difficult problem is expected from physical measurements and theoretical calculations [69,70].

## D. References


1. A. Krief: Tetrahedron *36*, 2531 (1980), and literature cited therein; D. Seebach: Angew. Chem. *91*, 259 (1979); Angew. Chem. Int. Edit. Engl. *19*, 239 (1979)
2. G. L. Closs, R. A. Moss: J. Am. Chem. Soc. *86*, 4042 (1964)
3. Köbrich et al.: Angew. Chem. *79*, 15 (1967); Angew. Chem. Int. Edit. Engl. *6*, 41 (1967); G. Köbrich: Angew. Chem. *84*, 557 (1972); Angew. Chem. Int. Edit. Engl. *11*, 473 (1972)
4. M. Jones Jr., R. A. Moss (eds.): Carbenes, Vols. I and II, Wiley, Interscі. Publ., New York 1973
5. R. M. Cory, F. R. McLaren: J. Chem. Soc., Chem. Commun. *1977*, 587
6. L. A. Paquette et al.: J. Am. Chem. Soc. *102*, 643 (1980)
7. H. Hopf: The preparation of allenes and cumulenes, in: The Chemistry of Ketenes, Allenes and Related Compounds, Vol. II, p. 779, S. Patai (ed.), Wiley, New York 1980

8. W. R. Moore, H. R. Ward, R. F. Merrit: J. Am. Chem. Soc. *83*, 2019 (1961)
9. R. M. Cory, L. P. J. Burton, R. G. Pecherle: Synth. Commun. *9*, 735 (1979)
10. M. S. Baird, P. S. Sadler: J. Chem. Soc., Chem. Commun. *1979*, 452
11. M. S. Baird, G. W. Baxter: J. Chem. Soc., Perkin Trans. *1*, 2317 (1979)
12. L. A. Paquette, E. Chamot, A. R. Browne: J. Am. Chem. Soc. *102*, 637 (1980)
13. G. Köbrich, H. Trapp: Z. Naturforsch. *18b*, 1125 (1963)
14. G. Cainelli et al.: Tetrahedron *27*, 6109 (1971); G. Cainelli, N. Tangari, A. U. Ronchi: Tetrahedron *28*, 3009 (1972)
15. J. Villieras, M. Rambaud: Synthesis *1980*, 644; J. Villieras et al.: Synthesis *1981*, 68
16. J. Villieras et al.: J. Organomet. Chem. *50*, C 7 (1973); J. Villieras, C. Bacquet, J. F. Normant: Bull. Soc. Chim. France *1975*, 1797
17. G. Köbrich: Chem. Ztg. *97*, 349 (1973)
18. J. Villieras, P. Perriot, J. F. Normant: Bull. Soc. Chim. France *1977*, 765; J. Villieras, P. Perriot, J. F. Normant: Synthesis *1979*, 968; H. Taguchi, H. Yamamoto, H. Nozaki: J. Am. Chem. Soc. *96*, 3010 (1974); H. Taguchi, H. Yamamoto, H. Nozaki: Bull. Chem. Soc. Jpn. *50*, 1588 (1977); J. Villieras, M. Rambaud: C. R. Acad. Sci. Paris C *290*, 295 (1980); J. Villieras, M. Rambaud: ibid. *291*, 105 (1980)
19. J. Villieras, C. Bacquet, J. F. Normant: Bull. Soc. Chim. France *1974*, 1731; J. Villieras, C. Bacquet, J. F. Normant: J. Organomet. Chem. *97*, 325 (1975); J. Villieras, C. Bacquet, J. F. Normant: ibid. *97*, 355 (1975); P. Entmayr, G. Köbrich: Chem. Ber. *109*, 2175 (1976)
20. P. Perriot, J. F. Normant, J. Villieras: C. R. Acad. Sci. Paris C *289*, 259 (1979)
21. H. Taguchi, H. Yamamoto, H. Nozaki: Bull. Chem. Soc. Jpn. *50*, 1592 (1977)
22. J. Villieras, P. Perriot, J. F. Normant: Synthesis *1979*, 502; C. J. Kowalski, K. W. Fields: J. Am. Chem. Soc. *104*, 321 (1982)
23. D. H. Williams et al.: Tetrahedron Lett. *1981*, 3745
24. J. Villieras et al.: Synthesis *1975*, 524; J. Villieras et al.: Synthesis *1975*, 533
25. L. W. Christensen, J. M. Seaman, W. E. Truce: J. Org. Chem. *38*, 2243 (1973)
26. V. Reutrakul et al.: Chem. Lett. *1979*, 209; V. Reutrakul, P. Thamnuson: Tetrahedron Lett. *1979*, 617; V. Reutrakul, W. Kanghae: Tetrahedron Lett. *1977*, 1225
27. V. Reutrakul, W. Kanghae, Tetrahedron Lett. *1977*, 1377
28. T. Durst et al.: Can J. Chem. *57*, 258 (1979)
29. D. F. Taber, B. P. Gunn: J. Org. Chem. *44*, 450 (1979)
30. K. M. More, J. Wemple: J. Org. Chem. *43*, 2713 (1978)
31. F. Cooke, P. Magnus: J. Chem. Soc., Chem. Commun. *1977*, 513; C. Burford et al.: J. Am. Chem. Soc. *99*, 4536 (1977)
32. P. Magnus, G. Roy: J. Chem. Soc., Chem. Commun. *1979*, 822
33. D. S. Matteson, D. Majumadar: J. Organomet. Chem. *184*, C 41 (1980); D. S. Matteson, D. Majumadar: J. Am. Chem. Soc. *102*, 7588 (1980)
34. D. Seyferth, J. L. Lefferts, R. L. Lambert: J. Organomet. Chem. *142*, 39 (1977)
35. G. L. Larson, O. Rosario: J. Organomet. Chem. *168*, 13 (1979)
36. J. Villieras et al.: J. Organomet. Chem. *190*, C 31 (1980)
37. P. Coutrot et al.: Synthesis *1977*, 615
38. J. Villiereas, A. Reliquet, J. F. Normant: Synthesis *1978*, 27
39. J. Villieras, P. Perriot, J. F. Normant: Synthesis *1978*, 29 (1978); J. Villieras, P. Perriot, J. F. Normant: Synthesis *1978*, 31
40. P. Perriot, J. Villieras, J. F. Normant: Synthesis *1978*, 33
41. P. Coutrot, P. Savignac: Synthesis *1978*, 34
42. P. Savignac et al.: Synthesis *1975*, 535; P. Savignac, P. Coutrot: Synthesis *1977*, 197
43. G. Köbrich, W. Goyert: Tetrahedron *24*, 4327 (1968)
44. D. Seyferth, R. L. Lambert, M. Massol: J. Organomet. Chem. *88*, 255 (1975); D. Seyferth, R. L. Lambert, J. Organomet. Chem. *88*, 287 (1975)
45. K. Kitatani, T. Hiyama, H. Nozaki: Bull. Chem. Soc. Jpn. *50*, 3288 (1977)
46. K. Kitatani, T. Hiyama, H. Nozaki: Bull. Chem. Soc. Jpn. *50*, 1600 (1977)
47. K. Kitatani et al.: Bull. Chem. Soc. Jpn. *50*, 2158 (1977)
48. D. Seebach: M. Braun, N. du Preez, Tetrahedron Lett. *1973*, 3509; M. Braun, D. Seebach; Chem. Ber. *109*, 669 (1976)

49. M. Braun, D. Seebach: Angew. Chem. *86*, 279 (1974); Angew. Chem. Int. Edit. Engl. *13*, 277 (1974); M. Braun, R. Dammann, D. Seebach: Chem. Ber. *108*, 2368 (1975)
50. R. Dammann, D. Seebach: Chem. Ber. *112*, 2167 (1979)
51. R. Hässig, H. Siegel, D. Seebach: Chem. Ber. *115*, 1990 (1982)
52. T. Hiyama et al.: Tetrahedron Lett *1978*, 3047
53. T. Hiyama et al., Tetrahedron Lett. *1979*, 2043
54. J. F. Normant et al.: Synthesis *1975*, 122
55. R. Sauvètre et al.: Synthesis *1978*, 128; C. Chuit et al.: J. Chem. Res. *1977*, 104
56. D. Masure et al.: Synthesis *1978*, 458
57. J. L. Hahnfeld, D. J. Burton: Tetrahedron Lett. 1975, 773
58. F. Karrenbrock, H. J. Schäfer: Tetrahedron Lett. *1979*, 2913
59. L. Duhamel, J. M. Poirier: J. Org. Chem. *44*, 3585 (1979)
60. G. Zweifel, W. Lewis, H. P. On: J. Am. Chem. Soc. *101*, 5101 (1979)
61. H. Siegel: Eidgenössische Technische Hochschule Zürich, unpublished results 1979
62. B. Mauzé: J. Organomet. Chem. *170*, 265 (1979); B. Mauzé: ibid. *202*, 233 (1980)
63. D. Seyferth, G. J. Murphy, B. Mauzé: J. Am. Chem. Soc. *99*, 5317 (1977)
64. T. Hiyama et al.: Bull. Chem. Soc. Jpn. *53*, 1010 (1980)
65. D. Seyferth, K. R. Wursthorn: J. Organomet. Chem. *182*, 455 (1979)
66. D. Seyferth et al.: J. Org. Chem. *45*, 2273 (1980)
67. L. Duhamel, J. Y. Valnot: Tetrahedron Lett. *1979*, 3319
68. E. Wenkert et al.: Synth. Commun. *9*, 11 (1979)
69. D. Seebach et al.: Angew. Chem. *91*, 844 (1979); Angew. Chem. Int. Edit. Engl. *18*, 784 (1979); H. Siegel, K. Hiltbrunner, D. Seebach: Angew. Chem. *91*, 845 (1979); Angew. Chem. Int. Edit. Engl. *18*, 785 (1979); D. Seebach et al.: Helv. Chim. Acta *63*, 2046 (1980)
70. T. Clark, P. v. R. Schleyer: J. Chem. Soc., Chem. Commun. *1979*, 833; T. Clark, P. v. R. Schleyer: Tetrahedron Lett. *1979*, 4963; T. Clark, P. v. R. Schleyer: J. Am. Chem. Soc. *101*, 7747 (1979); T. Clark, H. Körner, P. v. R. Schleyer: Tetrahedron Lett *1980*, 743

# Pyridinophanes, Pyridinocrowns, and Pyridinocryptands

**Veronica K. Majestic and George R. Newkome**

Department of Chemistry, Louisiana State University Baton Rouge, LA, 70803, USA

**Table of Contents**

**1 Introduction** . . . . . . . . . . 80

**2 Pyridinophanes** . . . . . . . . . . 81
2.1 2,6-Pyridinophanes . . . . . . . . . . 81
2.2 3,5-Pyridinophanes . . . . . . . . . . 87
2.3 2,5-Pyridinophanes . . . . . . . . . . 89
2.4 Complex Pyridinophanes . . . . . . . . . . 91

**3 Pyridinocrowns** . . . . . . . . . . 93
3.1 Coronands Containing Nitrogen Heteroatoms . . . . . . . . . . 93
3.1.1 Syntheses by the Schiff Base Condensation of Polyamines with 2,6-Diacetylpyridine or 2,6-Diformylpyridine . . . . . . . . . . 94
3.1.1.1 Syntheses of Coronands Containing Appendages Capable of Complexation . . . . . . . . . . 98
3.1.1.2 Syntheses of Coronands Containing the Triazine and/or Isoindolinylidene Subunits . . . . . . . . . . 99
3.1.1.3 Syntheses of Coronands Bridged by Hydrazine Moieties . . 101
3.1.2 Synthesis of Coronands by Condensation of Diacid Halides with Polyamines . . . . . . . . . . 103
3.1.3 Syntheses by the Nucleophilic Substitution of Dihalides by Diamines 104
3.2 Coronands Containing Oxygen and Nitrogen Heteroatoms . . . . . . 106
3.2.1 Synthesis by Schiff Base Condensation . . . . . . . . . . 106
3.2.2 Synthesis by Condensation of a Diacyl Halide and a Diaminoether 109
3.2.3 Synthesis by Nucleophilic Substitution of a Dihalide by an Amino Glycol . . . . . . . . . . 110

**4 Pyridinocryptands** . . . . . . . . . . 111
4.1 Synthesis by Condensation of Diacyl Halides and Diamines . . . . . . 111
4.2 Synthesis by Nucleophilic Substitution of Dihalides with Glycolates . . 113
4.3 Synthesis by Quaternization-Demethylation . . . . . . . . . . 114

**5 Acknowledgements** . . . . . . . . . . 115

**6 References** . . . . . . . . . . 115

In general, the pyridinophanes, pyridinocrowns, and pyridinocryptands have been surveyed. The pyridinophanes have been limited to those containing carbon bridging units of two atoms or less. In these instances the pyridinophanes may be bridged at the 2,6-, 3,5-, or 2,5-positions. The pyridinocrowns discussed contain carbon and nitrogen heteroatoms or nitrogen and oxygen heteroatoms and are bridged at only the 2,6-positions. The pyridinocrowns containing only oxygen or sulfur have not been included in this discussion. The discussion of the pyridinocryptands has not been restricted, since there are currently only a limited number of cryptands of this type.

## 1 Introduction

The current interest in heteromacrocycles and cryptands stems primarily from a desire to design ligands which can selectively complex cations. Since pyridine contains a nitrogen atom capable of complexation it has been of interest to study the effect of the pyridino subunit on complexation by incorporation of this moiety into two- and three-dimensional macrocyclic frameworks. A number of ill-defined factors influence cation selectivity and in order to design ligands to complex a particular cation these factors must be better understood. With regard to the ligand, there are several structural considerations: cavity size, nature and number of coordination sites, lipophilicity, and ligand conformation and flexibility. Other factors which influence cation selectivity and complex stability include the reaction environment (i.e., solvent polarity and temperature) and the counterion. Usually, a change from a polar solvent to a less polar one results in the formation of more stable complexes. As far as the counterion is concerned, nucleophilicity and steric bulk must be considered. Once these factors are weighed along with the charge density of the metal ion, it is possible to predict whether a particular complex will form and, if so, what the preferred stoichiometry will be.

This overview has been restricted due to the extent of this ever-expanding field. In general, the pyridinophanes, pyridinocrowns, and pyridinocryptands have been surveyed.

The pyridinophanes have been limited to those containing carbon bridging units of two atoms or less. In these instances the pyridinophanes may be bridged at the 2,6-, 3,5-, or 2,5-positions. These phanes are generally synthesized by one of several reaction pathways: pyrolytic extrusion of sulfur dioxide, photochemical extrusion of sulfur, Stevens rearrangement, Ramburg-Backlund rearrangement, or Hofmann-1,6-elimination.

The pyridinocrowns discussed contain carbon and nitrogen heteroatoms or nitrogen and oxygen heteroatoms and are bridged at only the 2,6-positions. The pyridinocrowns containing only oxygen or sulfur have not been included in this discussion. These pyridinocrowns have been synthesized by one of several reactions: Schiff-base condensation, condensation of a diacyl halide with a diamine, or nucleophilic substitution of a dihalide by a diamine.

The discussion of the pyridinocryptands has not been restricted, since there are currently only a limited number of cryptands of this type. The major synthetic pathways utilized for their construction have been the condensation of diacyl halides with diamines, nucleophilic substitution of dihalides by glycolates, and quaternization-demethylation of diamines.

# 2 Pyridinophanes

## 2.1 2,6-Pyridinophanes

In 1958, the synthesis of *2* was first reported [1]; the most satisfactory route was found to be the slow addition of butyllithium to a solution of 1,2-*bis*(6'-bromomethyl-2'-pyridyl)ethane (*1*) in ether to afford cyclophane *2* in 28% yield. The X-ray analysis shows *2* to possess a nonplanar chair conformation in the crystalline state [2]. Also, the $^1$H NMR spectrum of *2* in deuteriochloroform shows the carbon bridge methylene signal to coalesce at 13.5 °C and to exhibit an $A_2B_2$ pattern at −40 °C. The calculated barrier to inversion for *2* is 14.8 kcal/mole. The coupling constants of the low temperature spectra are also consistent with a chair-like conformation [3].

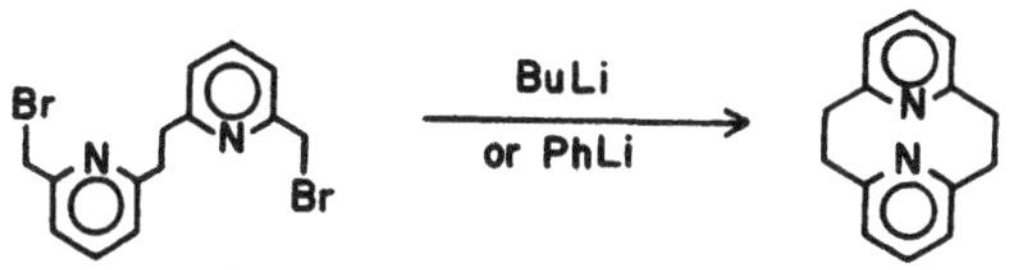


*1* *2*

Kauffmann et al. [4] reported the synthesis of *2* (1%) by the lithiation of 2,6-*bis*-(chloromethyl)pyridine and subsequent coupling in the presence of copper(II) chloride. The utilization of high dilution techniques did not affect the yields.

Pyridinophane *3* has also been synthesized in a four-step procedure which includes a pyrolytic sulfur dioxide extrusion as illustrated in Scheme 1 [5]. The UV photoelectron spectrum of *2* has been studied on the basis of a pertubational molecular orbital analysis and indicates the presence of through-space and through-bond interactions between the two pyridine rings. This finding has also been substantiated by CNDO/2 calculations. In addition, there appears to be a strong interaction between the nitrogen lone pairs and the π-system of the rings [6].

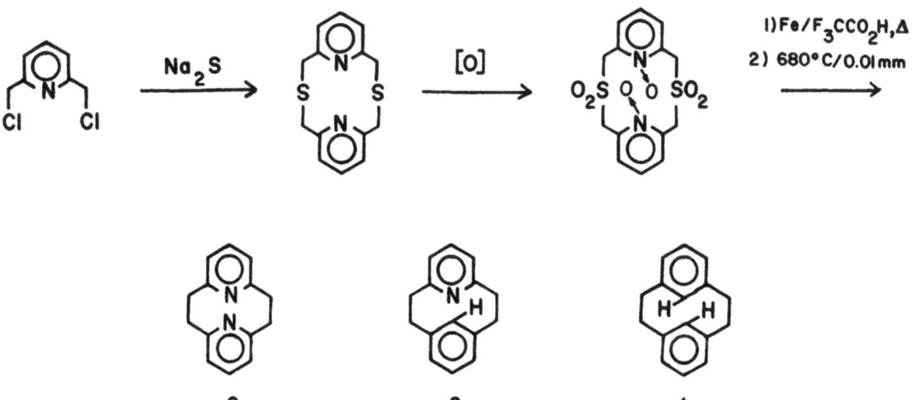


*2* *3* *4*

*Scheme I*

In an effort to evaluate conformational properties, pyridinophane *3* was synthesized [7–9] and compared to cyclophanes *2* and *4*. Perusal of the variable temperature NMR spectra of *2–4* demonstrates that *2* is the only cyclophane in this series which shows a temperature dependence: the $\Delta G^{*}$ for *2* has been found to be >27 kcal/mole. The $^{1}H$ NMR spectra of *3* and *4* remain unchanged to 200 °C; the bridging methylenes in *3* exhibit and ABCD pattern whereas in *4* and $A_2B_2$ pattern is observed. Comparison of these data with results reported by Vögtle [10] indicates the relative order of steric interactions (hydrogen-hydrogen, hydrogen-*N*-lone pair, and *N*-lone pair-*N*-lone pair) to be *2* < *3* < *4* [3,7].

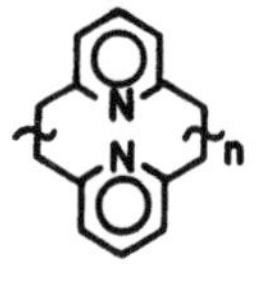

**5** n = 3

**6** n = 5

When 2,6-*bis*(bromomethyl)pyridine was treated with sodium in tetraphenylethylene, two macrocyclic ring systems were isolated: the trimer *5* (4.2%) and pentamer *6* (1.3%). In addition, the dimeric, trimeric, and tetrameric open-chain oligomers were isolated, each in <2% yield [11]. Tetramer *7* (4.2%), hexamer *8* (2.1%), and the octamer (<1%) were also obtained when 1,2-*bis*(6′-chloromethyl-2′-pyridyl)ethane was employed [12].

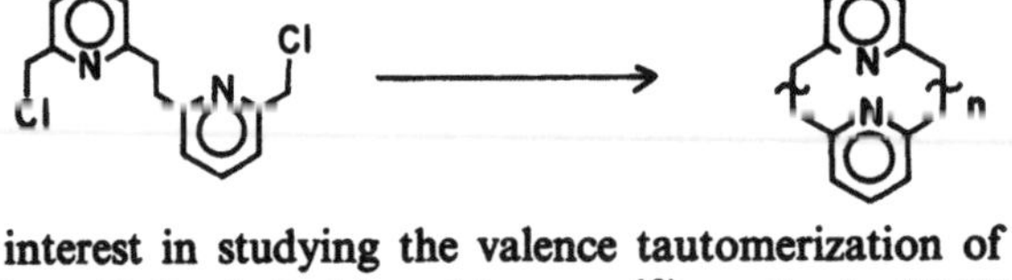

**7** n = 4

**8** n = 6

With an interest in studying the valence tautomerization of [2 · 2](2,6)pyridinophane-1,9-diene (*9*), Boekelheide and Lawson [13], synthesized *9* (20%) by the procedure indicated in Scheme II. The spectral data obtained for the isolated product

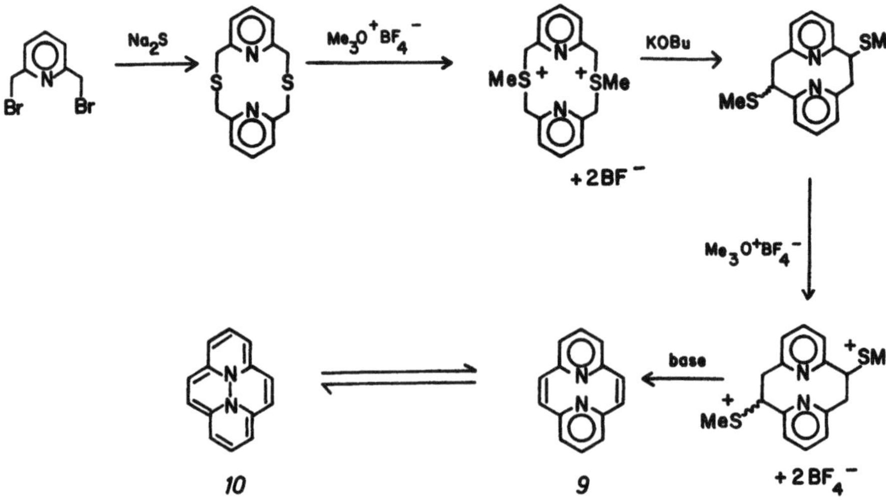


*Scheme II*

indicates that this cyclophane exists as *9* and not as its valence tautomer *10*. Studies also show that under aqueous acidic conditions *9* acts as a normal base and that there is no indication of an acid-catalyzed valence tautomerization to *10*.

Various substituted 2,6-pyridinophanes (*11–15*) have been synthesized in an attempt to force valence tautomerism by altering the bridge substitution [14]. Exchange of the bridge chloro groups with electron-donating and electron-withdrawing groups was expected to promote formation of the valence tautomer 2,6-pyridinophane by stabilization of the [14]-annulene form (*16* or *17*) over the cyclophane-diene form (*14* or *15*) by a "push-pull" through conjugation process [15]. Unfortunately, no evidence for valence tautomerism could be found for any of the compounds investigated. Attempts to introduce iodo, dimethylamino, nitro, or cyano groups by direct displacement reactions on *14* were also unsuccessful [16].

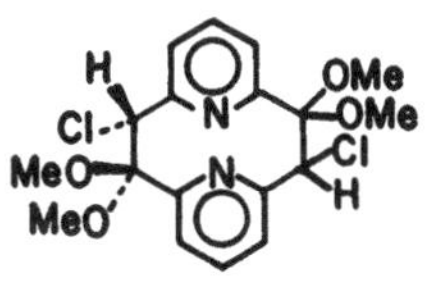


*11*

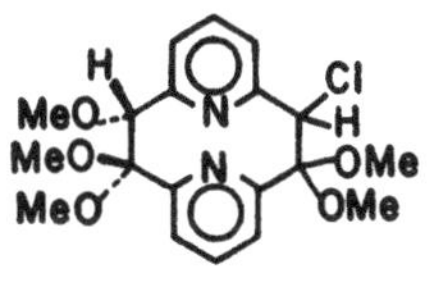


*12*

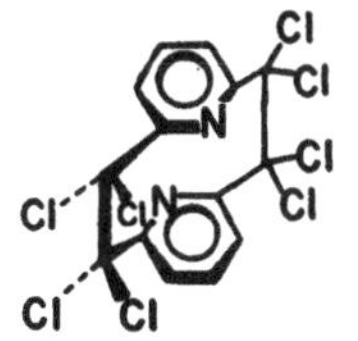


*13*

*14* R=Cl  *16* R=Cl

*15* R=SCH$_2$Ph  *17* R=SCH$_2$Ph

Cooke [17] has recently reported the synthesis of *18* in order to compare the steric interactions and rates of conformational flipping with other systems (*2*, *3*, *and 9*). The synthesis of *18* has been accomplished in two ways, both involving the intermediate 2-thiapyridinophane: one utilizes the Ramberg-Backlund rearrangement while the other depends on the Stevens rearrangement, as depicted in Scheme III.

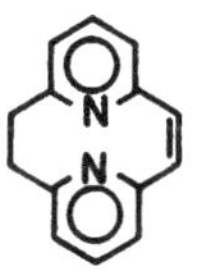


*18*

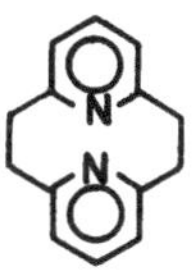


*2*

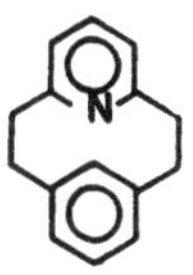


*3*

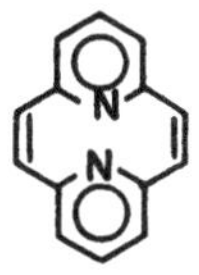


*9*

Variable temperature $^1$H NMR studies of *18* indicate a coalescence temperature for the methylenic protons at −43 °C, the frozen spectrum showing the expected AA'BB' pattern for the axial and equatorial protons on the methylene bridge at

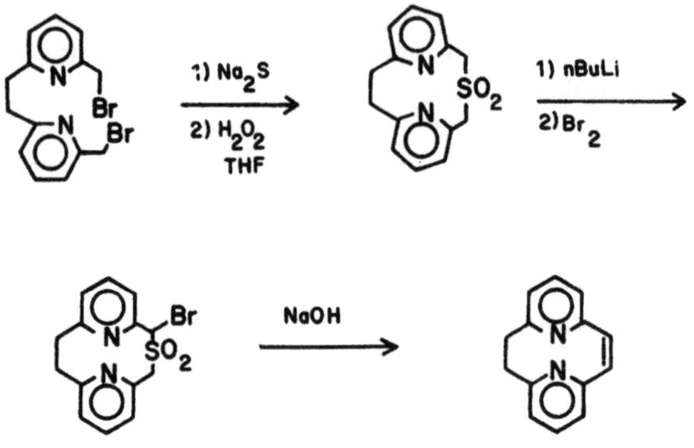


18

*Scheme III*

—65 °C. Thus, the calculated free energy of activation of *18* is about 10.9 kcal/mole, which is less than the value of 14.8 kcal/mole for *2*. It would be expected that pyridinophane *2*, containing two —$CH_2CH_2$— bridges, should be less mobile than *18* and should consequently have a higher energy of activation than *18*. However, for *9* the ethylenic protons appear as a singlet at 6.35 ppm and the spectra do not change over a wide temperature range (—80 °C to 100 °C), which supports a planar configuration and a single valence bond tautomeric structure [17].

Thiapyridinophanes *19* (51 %) and *20* (65 %) have been synthesized and characterized by normal physical and spectral data; however, the preparation of cyclophanes *21* and *22* by the classic photochemical extrusion of sulfur has not met with success [18].

19 21

20 22

In 1978, Haenel [19] reported the synthesis of pyridinophanes *23* (65 %) and *24* (44 %). According to variable temperature $^1$H NMR studies, the conformational barrier in *23* in 12.3 kcal/mole, a slightly higher value than that observed for *2*. The $^1$H NMR spectrum of *24* remains unchanged to —90 °C; this is consistent with either a rigid conformation, in which the pyridine ring is oriented perpendicular to the naphthalene ring, or with very rapidly interconverting conformers [19].

*23*

*24*

*27* *25*

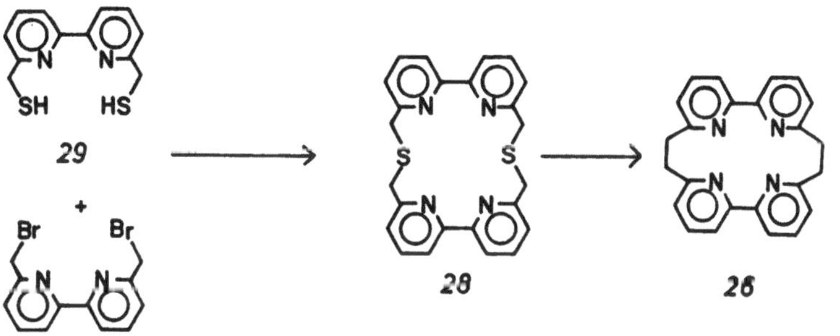

*29* *28* *26*

In this same vein, the synthesis of *25* [20] and *26* [21] has been attempted by initial formation of the cyclic sulfides, then extrusion of sulfur by irradiation in the presence of triethylphosphite. In both cases the limited sample size precluded formation of the desired cyclophanes (*25* and *26*) by this photolytic procedure. Variable temperature $^1$H NMR studies have been performed on *27* and *28*: *27* has a barrier to conformational inversion of 12.5 kcal/mole, while *28* has a corresponding barrier of 10.4 kcal/mole. Phane *28* exhibits a shift of the aromatic protons to higher field as compared to *29*; the most dramatic shift is observed for H-3, which absorbs at 7.72 ppm for *28* and at 8.37 ppm for *29*. This shift has been explained in terms of an approximately *syn* orientation of the pyridine rings in *28*.

In most studies on the conformational flipping of the derivatives of the [2 · 2]metacyclophanes, for which the critical steric interaction is due to the nature of the 8 and 16 positions, the order of increasing steric interaction is based on size. The interaction is minimized when both positions are occupied by nitrogen atoms and the lone-electron pairs interact. The interaction is increased when the *N*-lone pair of electrons interacts with a proton or larger substituent on the adjacent ring at the 16-position. Steric hindrance is maximized by the interaction of hydrogens attached to carbon atoms in the 8 and 16 positions. With this in mind, Boekelheide et al. [22] have synthesized *30* and *32* in order to study the steric interaction of the *N*-lone pair with the π-electron cloud of a *para*-bridged benzene ring. The synthetic methods employed include a Hofmann elimination of the sulfide and a pyrolytic sulfur extrusion.

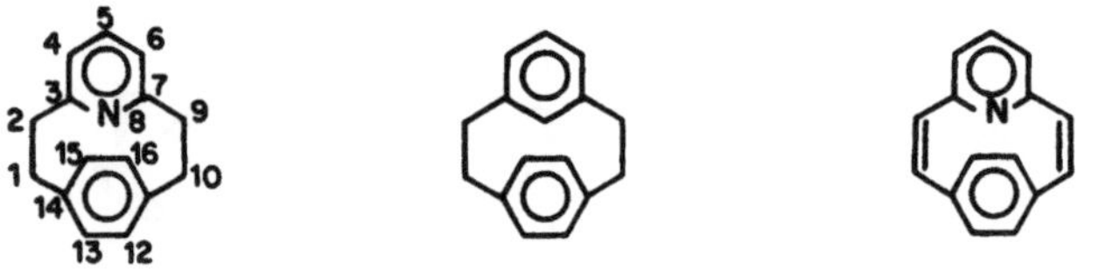

*30* (66%) *31* *32* (42%)

[2 · 2](2,6)-Pyridinoparacyclophane (*30*)[23] exhibits a temperature-dependent $^1$H NMR spectrum in which *30* is certainly undergoing conformational flipping. The $\Delta G^{\neq}$ value for *30* is 10.7 kcal/mole and is significantly smaller than the value of 20.6 kcal/mole for the carbocyclic analog *31*.

Phane *32* displays no temperature dependence in the $^1$H NMR spectrum; the single crystal X-ray analysis[24] of *32* shows the two aromatic rings exist in a near

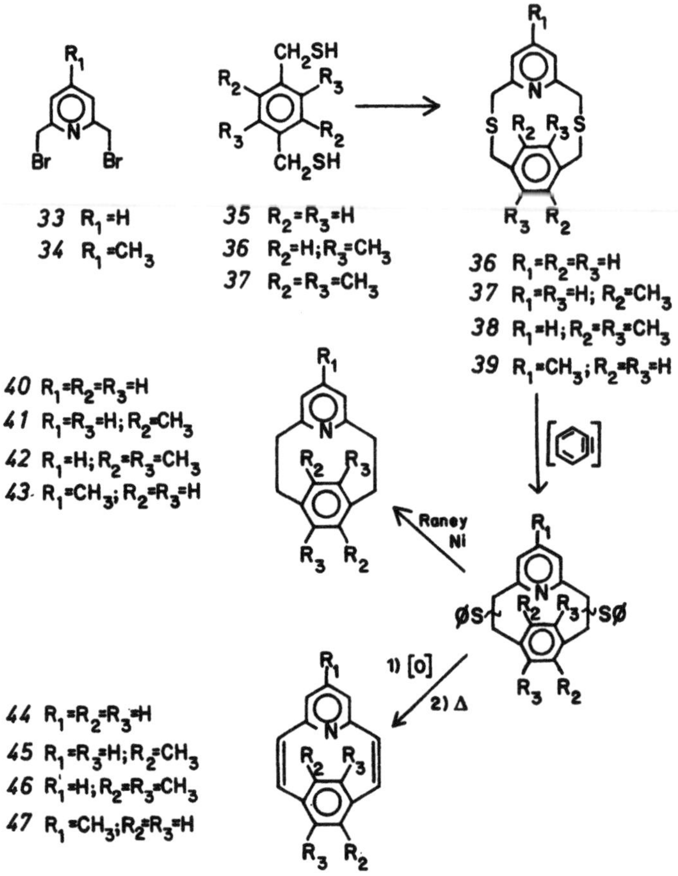


*Scheme IV*

perpendicular orientation. Whether or not the rings are perpendicular in solution is uncertain; however, the NMR suggests that the rings may be perpendicular in this instance [22].

So that a better understanding of the physical and chemical properties of these cyclophanes could be determined, the methyl-substituted pyridinoparacyclophanes *40–47* were synthesized (Scheme IV). The NMR studies performed on these cyclophanes indicate that the presence of the methyl groups on the *para*-bridged benzene moiety has very minimal effect on the conformational mobility.

It was expected that differences in the geometry between *40–42* and *44–47* would effect their relative basicities. Infrared spectroscopy [25,26] was selected as a means to measure this property and 2,6-dimethylpyridine ($pK_a$ 6.75) was chosen as a model. Cyclophane *40* exhibited the highest basicity ($pK_a$ 7.95) as a result of internal electron donation from the benzene ring to the pyridine ring within the macrocycle. Increased alkyl substitution (*41–43*) resulted in a decrease in basicity due presumably to increased steric hindrance. Cyclophane *42*, ($pK_a$ 7.38) possessing four methyl groups on the benzene moiety, has the lowest basicity for the unsaturated series. If the increased basicity recorded for *40–43* is due to electron donation from the benzoid system, an increase in the distance between the benzene ring and the pyridine ring should also decrease the basicity. This was found to be true for *36–38*, in which the rings are separated by three atoms.

In contrast, application of infrared techniques to *44–47* produced no measurable effects due to the much weaker basicity of these substances as compared to their saturated counterparts. This difference is attributed to variations in geometry between the unsaturated and saturated species. If the two rings are perpendicular to one another in the pyridinophane-1,9-diene system, as evidenced in the crystalline state [24], the *N*-lone pair is positioned in the center of the π-electron cloud of the benzene ring, and, therefore, is not accessible to a Lewis acid. The photoelectron spectra of these cyclophanes have also been studied and show nothing exceptional to indicate that they possess any unusual geometries [27].

## 2.2 3,5-Pyridinophanes

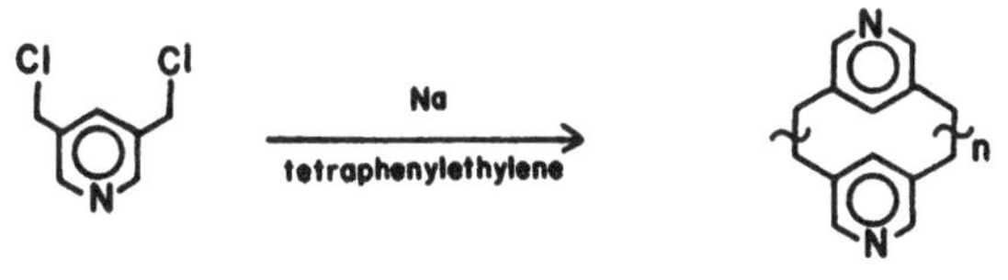


*49* n = 1
*50* n = 2
*51* n = 3

In 1967, the first 3,5-pyridinophane (*49*) was synthesized (2%) from 3,5-*bis*(chloromethyl)pyridine [28]. The $^1H$ NMR spectrum of *49* showed an AA'BB' pattern for the methylene protons at 3.22 and 2.18 ppm, analogous to the spectra of related carbophanes. Trimer *50* (4.2%) and tetramer *51* (1.5%) were also isolated from the reaction; however, the methylene signal in the $^1$H NMR spectrum of *50* and *51* consisted of a singlet at 2.83 and 2.77 ppm, respectively [29].

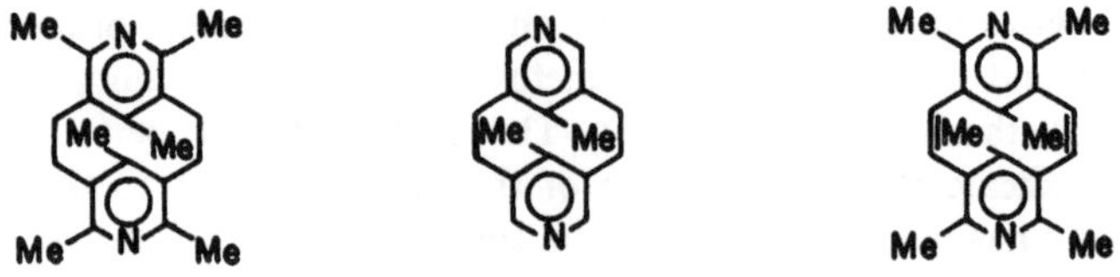

*52* *53* *54*

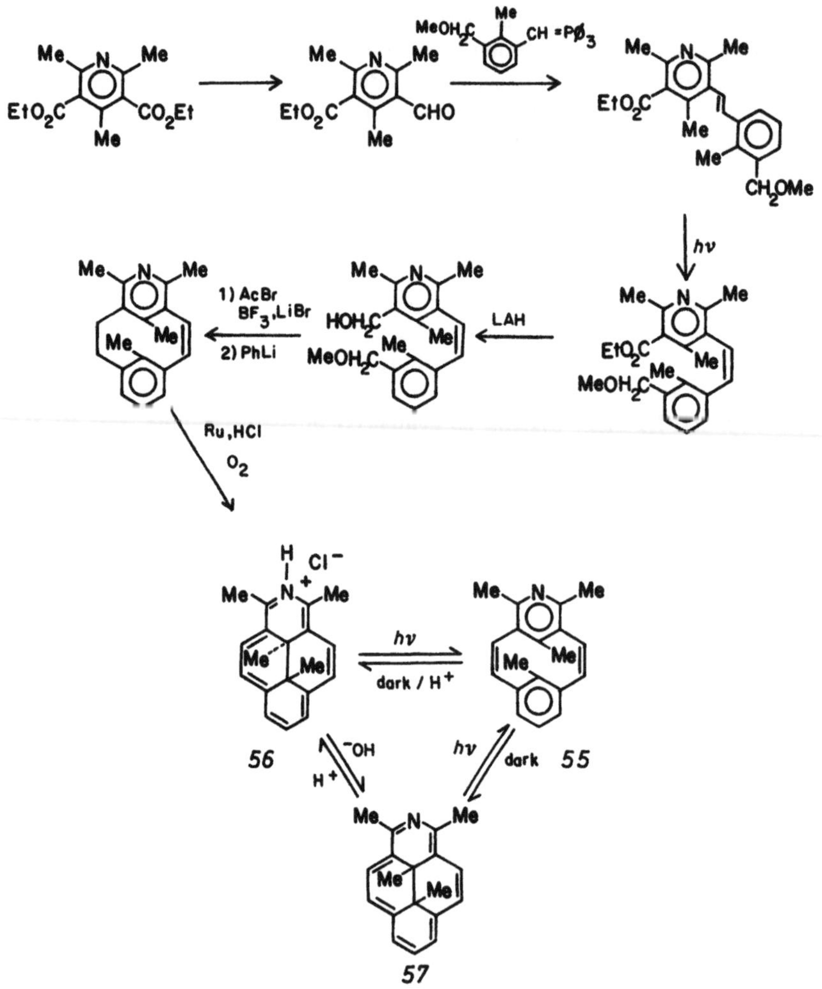

*57*

*Scheme V*

Other 3,5-pyridinophanes [*52* (15%) and *53* (15%)] have been synthesized and shown to be rigid-apparently in a chair-like conformation, according to the $^1$H NMR data [30].

The synthesis of *54* and *55* was accomplished by Boekelheide and Pepperdine in 1970[31]. Although metapyridinophane *54* was formed by a simple Wurtz coupling of the corresponding halogen derivative, *55* was best synthesized as shown in Scheme V. The most interesting feature of *55* and its corresponding hydrochloride salt (*56*) is that both are photochromic and undergo reversible valence tautomerization upon UV irradiation[31].

## 2.3 2,5-Pyridinophanes

The first 2,5-pyridinophanes *58* (60%) and *59* (23%) were synthesized by a photochemical extrusion of sulfur upon treatment with triethylphosphite[7,8].

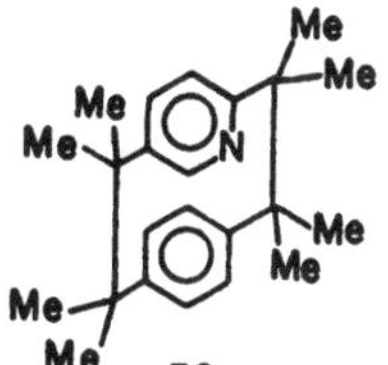


*58*

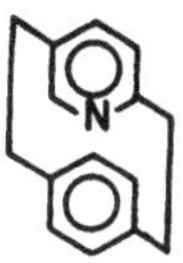


*59*

Wong and Paudler[32] were interested in preparing the mixed heterocyclophane *60* as the first example of a cyclophane containing both a π-deficient (pyridine) and a π-excessive (furan) ring. This was accomplished by a "cross" Hofmann degradation. To establish the correct structure for *60*, its UV and $^1H$ NMR spectra were compared with those of *61*, previously shown to exist in the staggered conformation. Both *60* and *61* displayed UV absorptions at $\lambda_{max}$ 224 nm and $\lambda_{max}$ 222 nm, respectively, indicative of no transannular interaction between the π-clouds of the furan ring and those of the pyridine ring. The $^1H$ NMR spectrum of *60* suggested that the most likely configuration is when the furan oxygen atom exists in closer proximity to the $N_1—C_6$ bond than to the $C_4—C_5$ bond.

Variable temperature $^1H$ NMR data obtained showed no significant spectral changes over a wide temperature range (—50 °C to 110 °C).

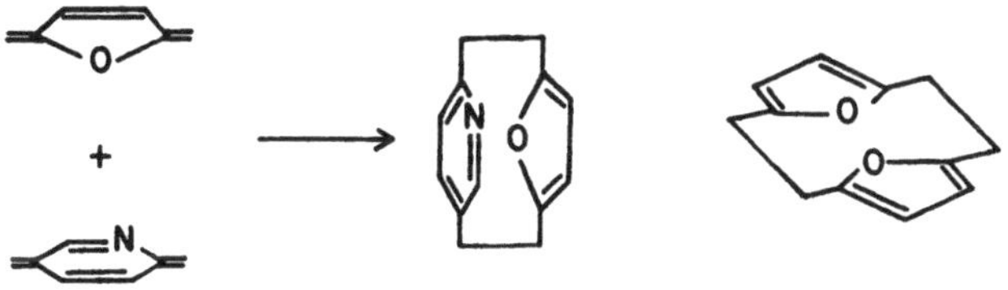


*60* (6.3%) *61*

Four isomers of the 2,5-pyridinophanes have also been synthesized (>50%) by the Hofmann degradation of *62* and were separated by gas chromatography. The four isomers were assigned the structures *63* (17%), *64* (24%), *65* (14.2%), and *66* (5.5%) on the basis of NMR data. (These percentages were calculated by VPC data.)[33,34]

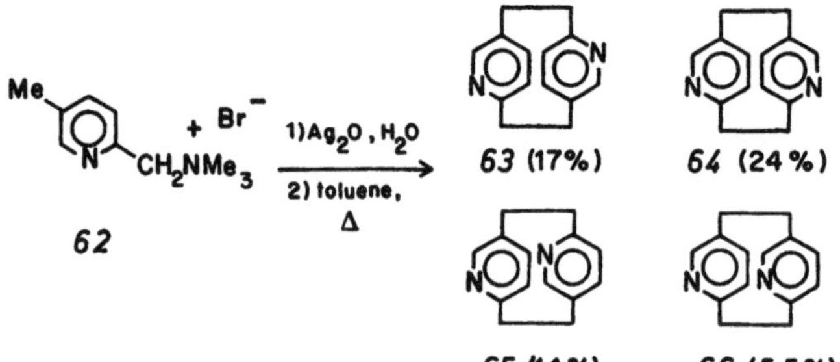


65 (14%) 66 (5.5%)

Modification of this reaction procedure to incorporate a benzene ring in place of one pyridine ring was accomplished by a "cross" Hofmann-1,6-elimination from an intimate mixture of *67* and *68*. Unfortunately, the yield of the desired cyclophane *69* was obtained in only 0.4% yield [35].

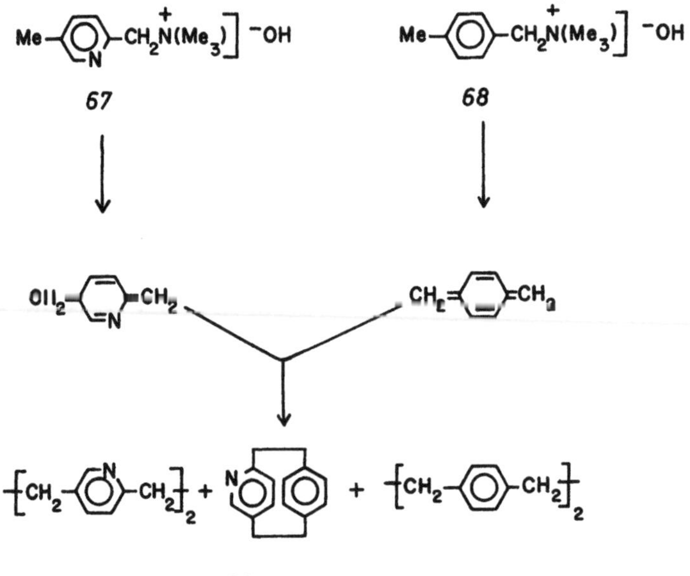


0.6% *69* (0.4%) 42%

Pyridinophanes *71* and *73* have been synthesized and their $^1H$ NMR spectra at room temperature exhibit an $A_2B_2$ pattern for the methylenic protons [36].

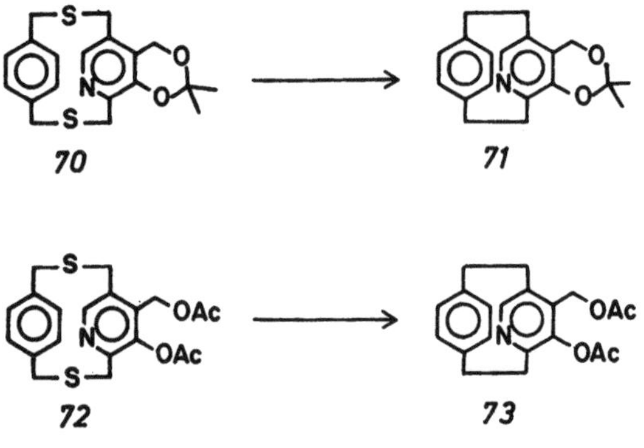

## 2.4 Complex Pyridinophanes

Several pyridinophanes have been synthesized in which the two pyridine rings are stacked one-on-top-of the other and are held together with four bridges. Cyclophanes *74* (24%) and *75* (22%) have been synthesized as shown in Scheme VI, and the single crystal X-ray analyses [37, 38] indicate that the heteroaromatic rings exist in the boat conformation with an inter-ring distance in both cases of 2.64 Å. Both cyclophanes are relatively strong bases and should be able to complex transition metals; to date no such studies have been published [39].

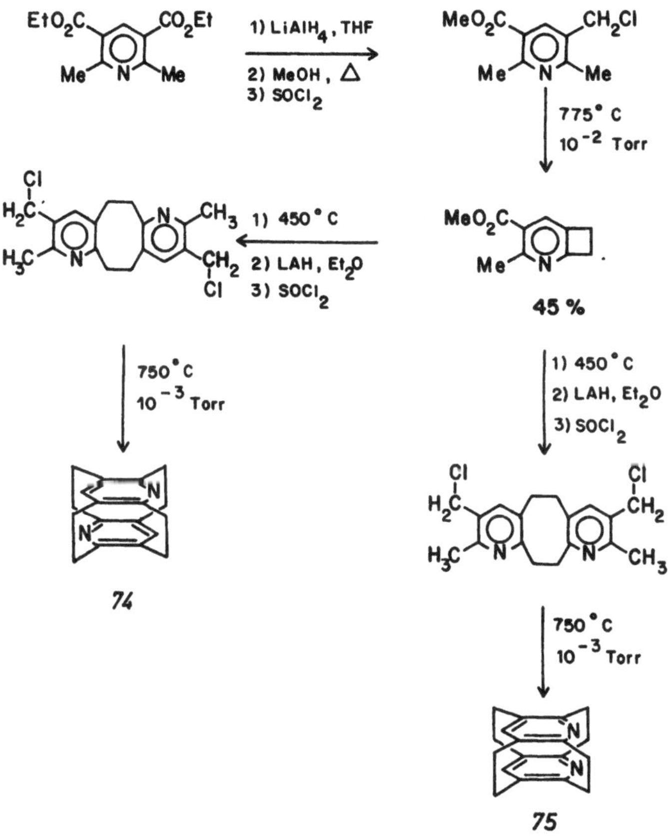


*74*

*75*

*Scheme VI*

The most complex pyridinophanes reported thus far are the stacked cyclophanes *76*, *77*, (Scheme VII) and *78* (Scheme VIII). The variable temperature $^1$H NMR spectra of *76* indicate that the molecule undergoes a conformational inversion with a barrier of 11 kcal/mole, comparable to that of several previously considered cyclophanes (eg., *31*, *40*, *41*, and *43*) [27].

The triply layered cyclophane diene *77* has both pyridine rings oriented perpendicular to the central benzene ring, according to single crystal X-ray determination. The central benzene moiety exists in a twist boat conformation and the pyridine

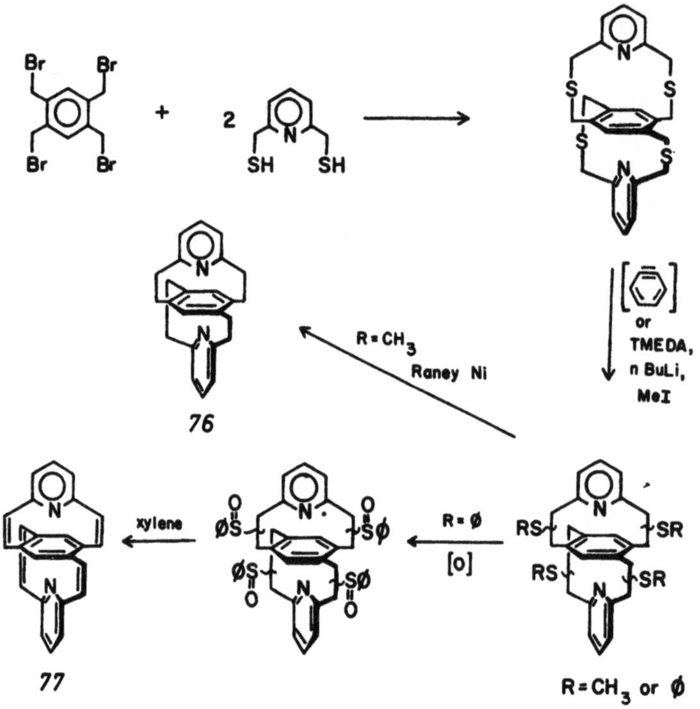


76

77

*Scheme VII*

rings deviate from a perpendicular arrangement by 15°; each nitrogen atom lies 2.52 Å[1] from the mean plane of the benzene centroid. No X-ray structure determination has been performed on *76*, thus the preferred conformation in the crystalline state is not yet known[40)].

In general, the pyridinophanes discussed are highly strained systems and relief of the inherent strain may be achieved by distortion of the bridging carbon atoms, puckering of the aromatic rings, or orientation of the aromatic rings in a non-parallel manner. X-ray crystal structure analyses have provided some insight into the importance of these factors[24, 32)]. However, in most instances the stresses and strains in these molecules are evenly spread throughout the system. In addition, the rates of conformational rotation have been performed by variable temperature NMR techniques and indicate that the sensitivity of the rate of conformational flipping is generally due to nonbonded interactions or changes in bond angles. For some pyridinophanes it is not clear whether the structure is conformationally rigid or rapidly flipping. Further studies must be performed on these systems so that they may be fully understood. There is yet much to be learned about the pyridinophanes including their conformational preferences, chemical properties, and complexation properties, if any.

---

[1] 1 Å = $10^{-8}$ cm

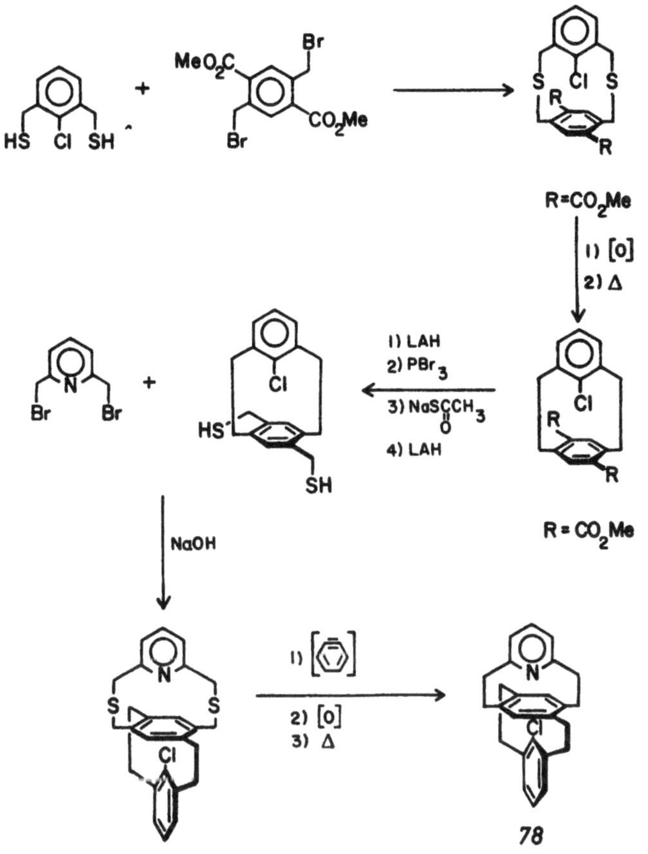


78

*Scheme VIII*

# 3 Pyridinocrowns

## 3.1 Coronands Containing Nitrogen Heteroatoms

Myriad polydentate aza-macrocycles have been reported [41]. The extent of the subject forces limitation of this discussion to only macrocycles containing a pyridine or dipyridine subunit. Most of these coronands have been synthesized by a Schiff base condensation of an aldehyde or ketone with a *bis*-primary amine in the presence of a metal ion. The metal ion acts as a template, resulting in dramatic increases in yield of the desired cyclic product over linear polymerization products [42–46]. Lindoy and Busch [45] have described this effect in two ways, kinetic and thermodynamic. If the metal ion controls the steric course of a series of stepwise reactions, the template effect is considered to be kinetic. If the metal ion influences an equilibrium in an organic reaction sequence by coordination with one of the reactants, the template effect is termed thermodynamic. It is the kinetic effect that is believed to be operative in most metal ion-assisted (*in situ*) syntheses of macrocycles, although, it is difficult to ascertain which of the two effects is dominant in a specific synthesis.

As in most synthetic schemes, there are advantages and disadvantages to these *in situ* macrocycle syntheses. The advantages include increased yields by virtue of the elimination or reduction of side reactions and macrocyclic product control by steric influence of the metal ion. The disadvantages include the fact that the steric requirements of the metal ion may preclude formation of the desired macrocyclic product and the occasional inability to remove the metal ion from the resulting complex due to the inherent instablity of the free ligand.

Some of the important factors influencing these reactions are: polarization effects, kinetic liability, thermodynamic stability, and stereochemistry of both the metal atom(s) and the reactants. In general, these syntheses are more metal ion specific than are complexations of the metal ion with a preformed macrocyclic ligand [45].

### 3.1.1 Syntheses by the Schiff Base Condensation of Polyamines with 2,6-Diacetylpyridine or 2,6-Diformylpyridine

The *in situ* synthesis of macrocyclic ligands by condensation of 2,6-diacetylpyridine with polyamines in the presence of metal ions represents one of the early demonstrations of the "template effect". Curry and Busch reported the first penta- and hexadentate macrocycles (*79* and *80*) [47], both of which are cyclized in the presence of iron(II) chloride tetrahydrate. The two ligands were isolated as highly

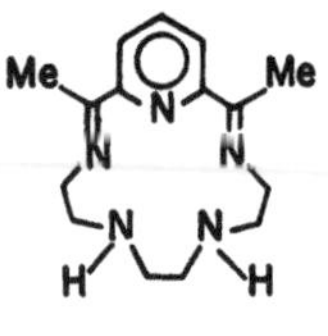

*79*

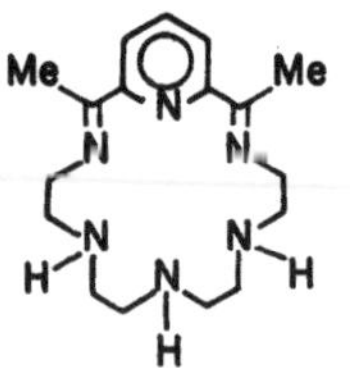

*80*

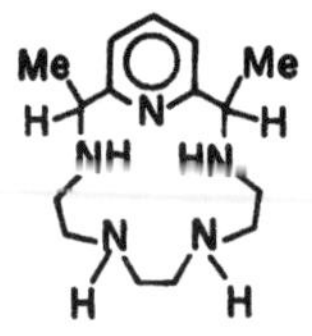

*81*

crystalline, spin-paired, iron(III) complexes, and were characterized by magnetic, conductive, and spectral measurements [47,52]. Because these complexes cannot be protonated, it is believed that all of the nitrogen atoms in the ligand are coordinated to the metal atom [49].

Other metals for which complexes of ligand *79* have been isolated are: iron(II) [48,55,56,58,59], iron(III) [47,52,54,56,59], zinc(II) [57–59], cadmium(II) [59,60,78], mercury(II) [58,78], magnesium(II) [59,60], manganese(II) [58], nickel(II) [79], and lead(II) [80]. The reduction of *bis*-amine *79* has been reported to afford ligand *81*. In addition, the transition metal complexes of iron(III), cobalt(III), nickel(II), and copper(II) with ligand *81* have been prepared and characterized [53].

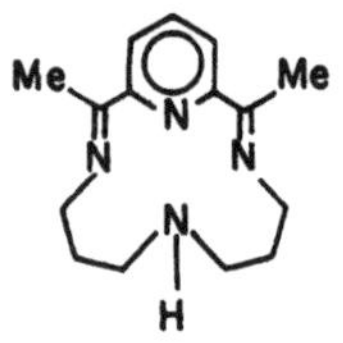

*82*

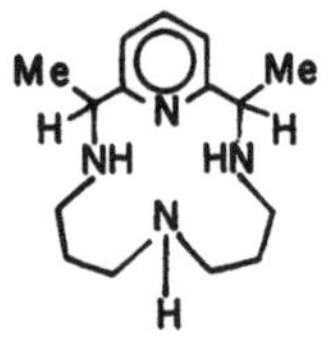

*83*

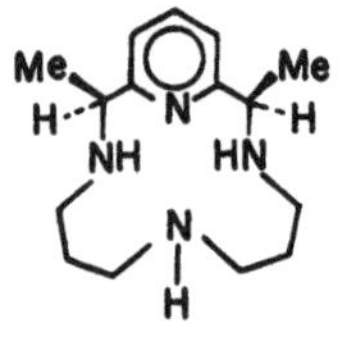

*84*

Complexes of *82* have also been formed by the reaction of 2,6-diacetylpyridine and *N,N-bis*(3-aminopropyl)amine in the presence of nickel(II) chloride and copper(II) chloride [50]. Other metals that have been used include: copper(II) [63,64], nickel [64,65], cobalt(II) [66], manganese(II) [73], cobalt(I) [69], cobalt(III) [68,70–72], zinc(II) [73], and ruthenium(II) [74]. Karn and Busch [51] have reported the catalytic hydrogenation of the nickel(II) perchlorate complex of *82* to afford two nickel(II) complexes of *83*: a yellow minor component and a red major component which preliminary studies indicate to be the meso form (*84*). The isomeric ligands can be displaced from the respective reduced complexes by cyanide ion. Ligand *84* has also been isolated and characterized as the cobalt(III) [67], iron(II) [61,62], iron(III) [62], and copper(II) complex [75,76]. Dehydro — *82* has also been synthesized and complexed with nickel(II) [65,65a], and nickel(III) [65a].

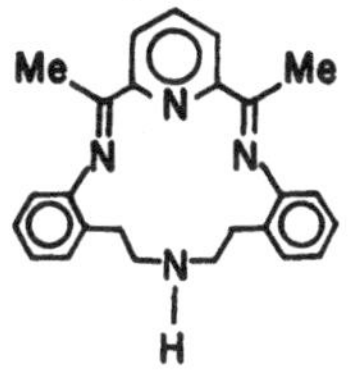


*85*

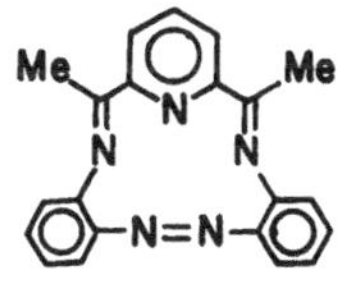


*86*

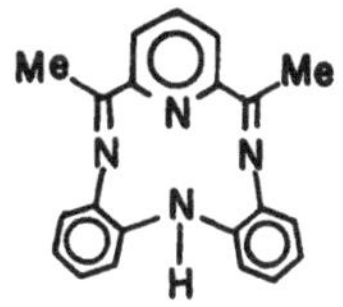


*87*

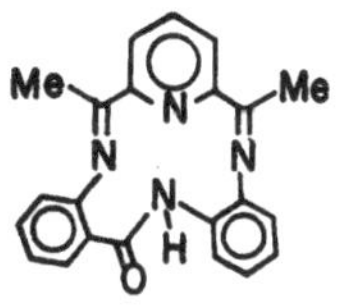


*88*

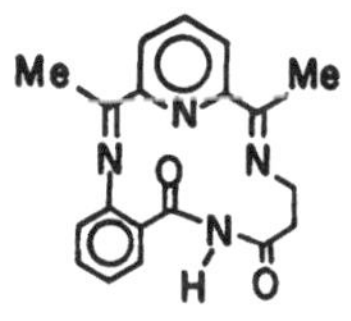


*89*

Treatment of other amines with 2,6-diacetylpyridine in the presence of nickel(II), copper(II), or cobalt(II) salts has generated a wide variety of new macrocyclic complexes (*85*–*89*). It is interesting to note that *87* forms complexes with nickel(II) and copper(II) but not with cobalt(II); no explanation for this observation has been forwarded [85].

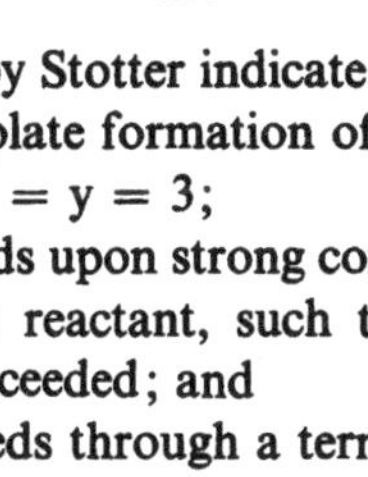


*90*

Studies by Stotter indicate that:
1. the template formation of macrocyclic complexes of *90* requires a minimum ring size of x = y = 3;
2. it depends upon strong complexation of the metal at the pH of the reaction by the triamine reactant, such that the solubility product of the metal hydroxide is never exceeded; and
3. it proceeds through a ternary intermediate complex.

If in *90* x = 3 and y = 4, the yields are lower and the complexes are less stable due to the presence of a seven-membered chelate ring in the complex. In contrast, when complexes containing five- or six-membered chelate rings are formed the yields are usually much higher, a result of the greater stability of such systems. Since no reaction occurs when the metal ion is present as a suspension of its hydroxide (and thus is not coordinated by the triamine) it is believed that the reaction between the 3,3-triamine and 2,6-diacetylpyridine is a true template reaction in which both reactants are coordinated to the metal ion [77].

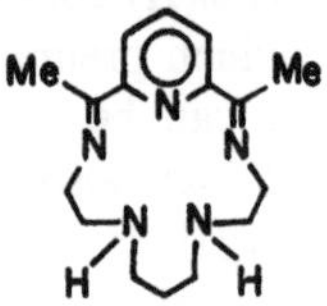

*91*

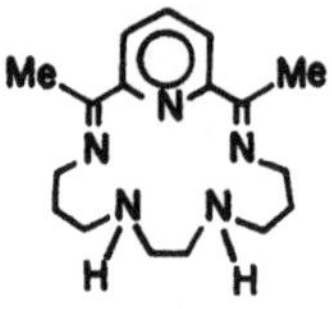

*92*

Complexes of macrocycles *91* [81] and *92* have been prepared *in situ* from 2,6-diacetylpyridine and the necessary diamine. Transition metal complexes of *91* with iron(II) [48,56,59,83], iron(III) [59,81], magnesium(II) [60], cadmium(II) [59,78,82], mercury(II) [78,82], silver(I) [84], maganese(II) [58,59,73,83], manganese(III) [73], nickel(II) [79], and zinc(II) [59,73,83], have been isolated and characterized. The isolated complexes of *92* include those of silver(I) [82,84], iron(II) [59], iron(III) [59], manganese(II) [58,59,73], manganese(III) [73], zinc(II) [59,73], cadmium(II) [59,78,82], mercury(II) [78,82], nickel(II) [79], lead(II) [80,86], magnesium, calcium, strontium, and barium [87].

The complexes of ligands *79* and *91* possess structures that are approximately pentagonal bipyramidal with the macrocycle defining the equatorial plane and the axial positions being occupied by a unidentate anion or water. The complexes of *92* are somewhat different, in that the five donor nitrogen atoms do not lie in one plane: there is a distortion which causes one face of the macrocycle to be more sterically crowded that the other [59,78,79]. The coordination number ranges from five to seven depending on the occupancy of the two axial positions.

The nickel(II) complex of *92* cannot be prepared directly *via* the template method, but can be prepared by a transmetallation procedure. Synthesis of the macrocycle in the presence of one of the metal ions known to be effective as a template is followed by a metal exchange process in solution to insert the nickel(II) ion. This cation exhibits a strong preference for the square planar, square pyramidal, and octahedral geometries [79]. Thus the failure of the nickel(II) cation to behave as a template ion in the synthesis of *92* is probably due to the disinclination of the metal to accommodate the pentagonal array of donor nitrogen atoms necessary for reaction to occur.

Attempts to prepare the nickel(II) complex of *91* by transmetallation have resulted instead in the formation of addition products *91a* or the ring-opened product *93*. In this case, the stereochemical preferences of the metal impose a new conformation on the ligand and thereby enhance its reactivity [79].

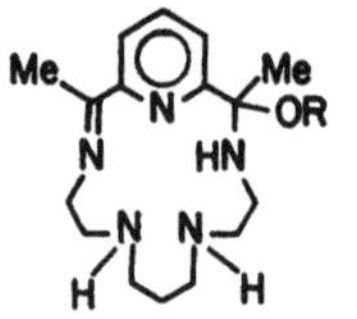

*91a* (R=Me, Et)

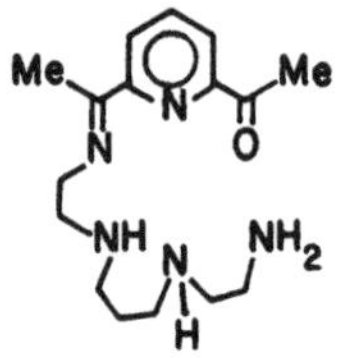

*93*

Judging from molecular models, the size of ligand *91* is not sufficient for its complexes to possess a planar conformation without experiencing a great deal of strain. The strong preference of nickel(II) for an octahedral geometry provides the driving force for addition across one of the azomethine bonds, which in turn introduces the flexibility necessary for achievement of the optimum conformation. It is important to point out that pentagonal bipyramidal complexes of *91* with other metal ions of similar size and charge, such as iron(II), are relatively stable to nucleophilic attack by alcohols or water [59,82]. The opposite effect has also been well documented; the steric requirements of the ligand may impose an unusual or irregular coordination geometry on a metal ion [59,82,88,89].

If 2,6-diacetylpyridine is condensed with *N,N-bis*(3-aminopropyl)-amine in the presence of a large metal ion such as silver, dimer *94* is the only observed product and exists as the disilver(I) complex. The Ag———Ag distance is 6.0 Å: clearly, there is no interaction between the metal atoms. It appears that in this case the metal ion size is the dominant factor in determining the course of the reaction [90]. Thus far, *94* is the only dimeric complex synthesized by the *in situ* template procedure.

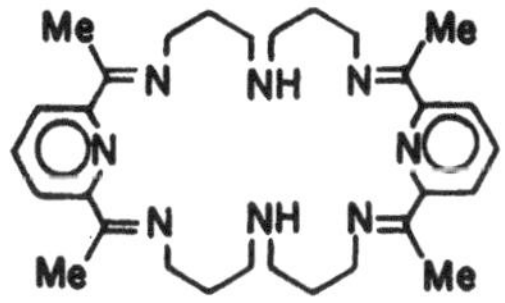

*94*

Stotz and Stoufer [91] reported the first example of a binuclear complex of a macrocyclic ligand *95*, which holds the two metal ions in sufficiently close proximity to permit metal-metal interactions, as indicated by magnetic susceptibility and esr measurements. The binuclear complex of ligand *95* (40—45%) is formed by the copper(II) assisted Schiff base condensation of 2,6-diacetylpyridine with *o*-phenylenediamine. Although calcium, strontium, barium, lead(II), lanthanum(III) [92,93], and cerium(III) [92] have all been successfully used as templating ions in the synthesis of ligand *96*, attempts to effect cyclization in the presence of magnesium, manganese(II), iron(II), nickel(II), copper(II), silver(I), cadmium(II), and mercury(II) have failed. These complexes of ligand *96* appear to be the first in which the "hard" alkaline

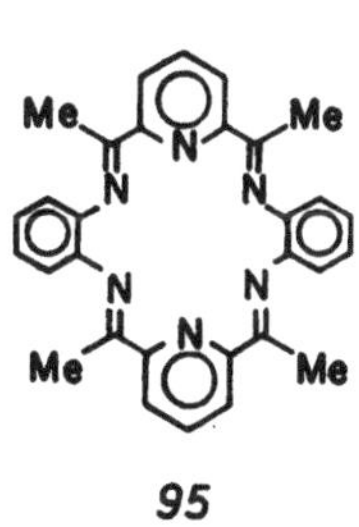

*95*

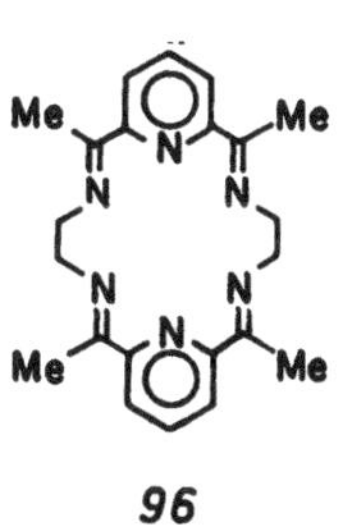

*96*

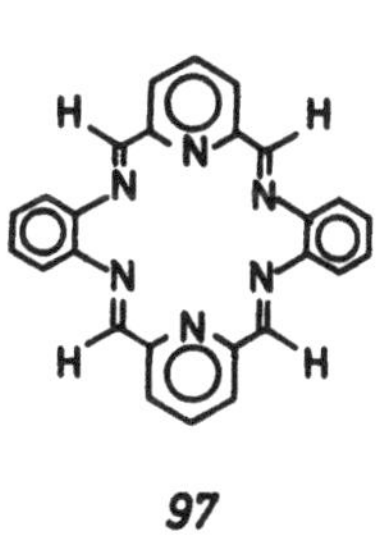

*97*

earth metal ions are bound to a macrocyclic ligand containing only "soft" nitrogen donors [94]. Efforts to prepare a barium complex of *95* have failed, whereas calcium, strontium, barium, and lead(II) have been used successfully in the synthesis of *97* by the Schiff base condensation of 2,6-diformylpyridine with *o*-phenylenediamine[95].

Transmetallation reactions in methanol, intended to form the manganese(II) (ionic diameter; 1.84 Å), iron(II) (1.84 Å), cobalt(II) (1.80 Å), or zinc(I) (1.80 Å) complexes of *97* (cavity radius; 2.7 Å), have led to the isolation of a new series of complexes where the ligand has undergone a ring contraction to better accomodate the smaller cations. The ring contraction was initiated by addition of methanol across one of the azomethine bonds during transmetallation, and subsequent formation of a five-membered imidazoline ring to afford ligand *98*. Evidence to support the belief that the ring contraction is due to a disparity in the sizes of the metal ion and ligand *97* is provided by the fact that exchange of cadmium(II) (2.20 Å) for barium(II) does not invoke ring contraction, but forms the desired complex[96].

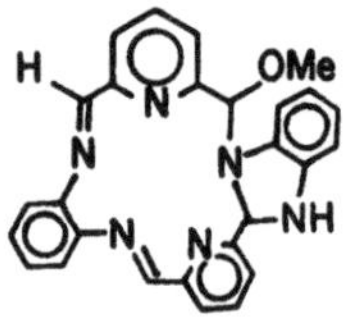

*98*

In an analogous fashion complex *100* may be obtained by condensation of 2,6-diformylpyridine with *99* in the presence of manganese(II) or zinc(II). According to crystallographic data, the coordination geometry of *100* is distorted pentagonal bipyramidal[97].

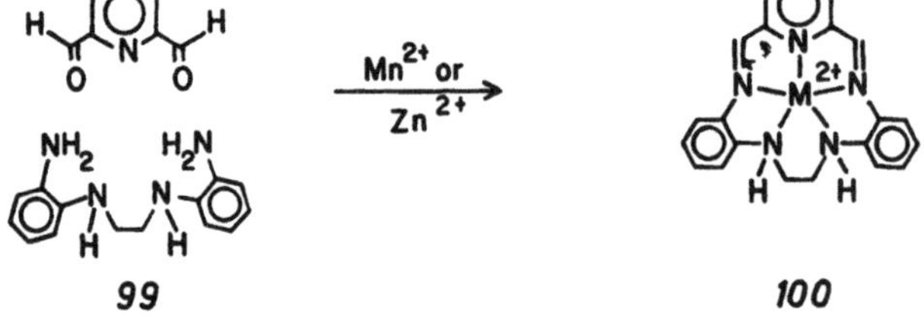

*99* *100*

### *3.1.1.1 Syntheses of Coronands Containing Appendages Capable of Complexation*

Several derivatives of *82* have been synthesized in which an additional functional group has been attached to the secondary $sp^3$ nitrogen atom. Such coronands are less rigid in terms of the geometrical requirements of the coordinated metal ion and are thus especially interesting for structural studies. Condensation of *N,N-bis*(3-aminopropyl)-*N′,N′*-dimethylethylenediamine with 2,6-diacetylpyridine in the presence of nickel(II) ions results in the formation of ligand *101*, which upon subsequent reduction of the imine groups affords the nickel(II) complexes of *102*[98]. Ligands *103* and *104*, in which the terminal amino groups have been replaced by hydroxide, have also been synthesized as the nickel(II) complexes. Ligands *101–104* may exist in one of two forms depending on the pH of the solution. At low pH protonation of the dimethylamino or hydroxy groups results in binding of the metal ion by the four nitrogen atoms of the macrocycle in a square planar ligand field. At

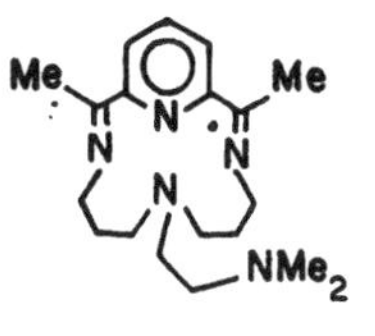
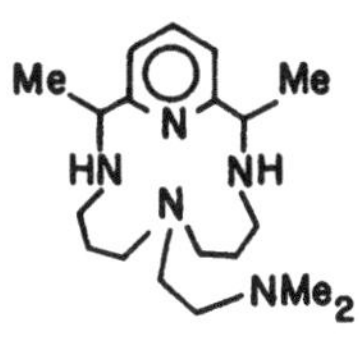
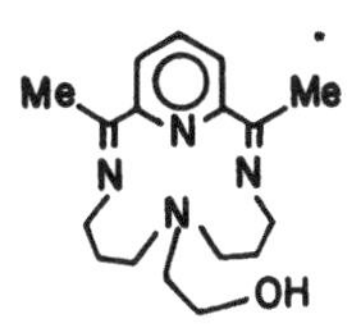
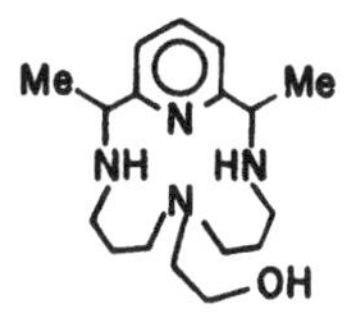

*101* *102* *103* *104*

higher pH, the dimethylamino or hydroxy groups can coordinate in one of the axial positions to produce a pseudo-octahedral coordination geometry [99].

Interest in these ligands stems from a desire to synthesize improved analogues of the cobalamines [100]. With this in mind, 2,6-diacetylpyridine was condensed with *N,N,N-tris*(aminopropyl)amine in the presence of nickel(II) or copper(II) to afford the complex of *105*. Cobalt(II) and zinc(II) have also been employed as templating agents in the synthesis of *105* [101]. The reaction of the nickel(II) complex of *105* with acetone results in a dimeric complex, *106* [101], by a process that is well established for primary amines [102].

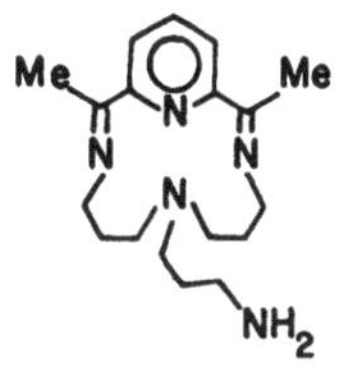

*105*

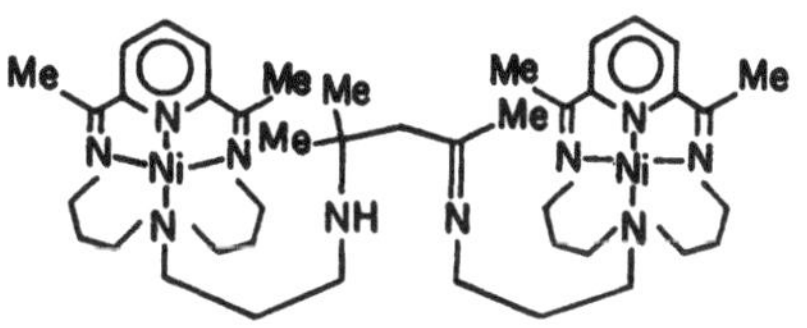

*106*

#### 3.1.1.2 *Syntheses of Coronands Containing The Triazine and/or Isoindolinylidene Subunits*

Borodkin et al. have reported the syntheses of a series of macrocycles containing isoindoline and triazine subunits with 2,6-diaminopyridine (*110*) [103–108]. The first macrocycle of this type reported involves the treatment of 4-chloro-2,6-*bis*[(1-imino-3-isoindolinylidine)amino]triazine with 2,6-diaminopyridine to afford *107* (25%).

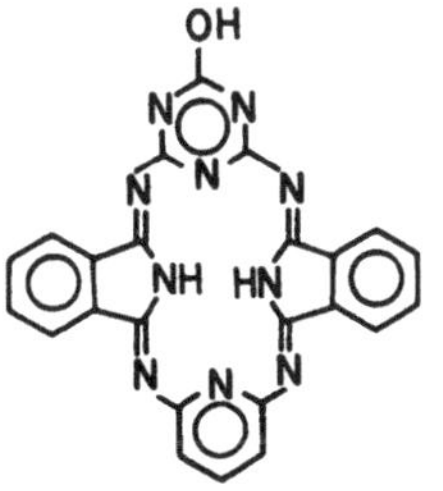

*107* (25%)

Complexes of *107* have been isolated for copper(II), cobalt(II), and nickel(II) cations [103]. Condensation of 2,6-diaminopyridine with *108* affords the dimeric macrocycle *111* (70%). No complexes have been reported for ligands *111* [108] or *113* [104,106]. In addition, the reaction of 1,3-*bis*[(1-imino-4,5,6,7-tetrahydro-3-isoindoline)amino]benzene with 2,6-diaminopyridine gives *113* (70%) [104,107].

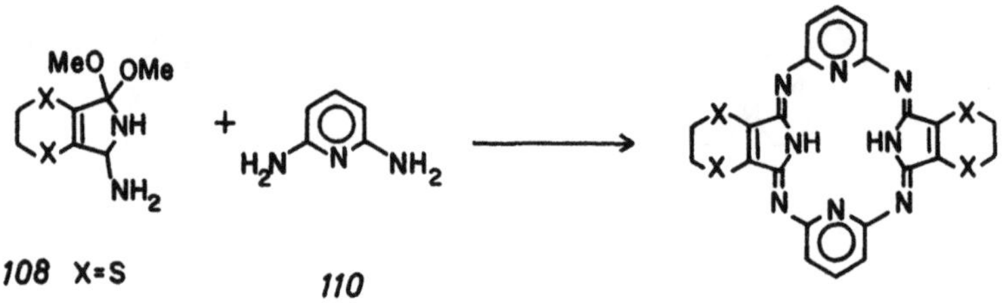

*108* X=S

*109* X=$CH_2$

*110*

*111* X=S

*112* X=$CH_2$

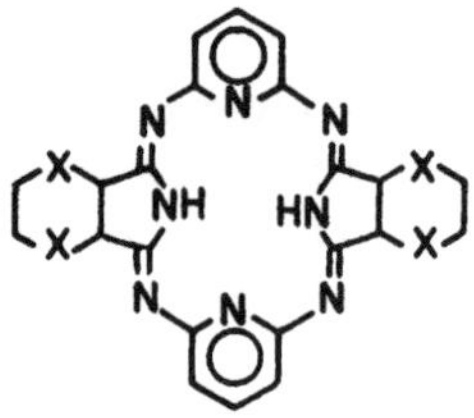

*113* X=$CH_2$

*114* X=S

These authors have also reported the preparation of a series of chlorotriazines *115–118* with 2,6-diaminopyridine to generate the corresponding dimeric macrocycles *119–122*. The copper(II) complexes of these macrocycles (*119–122*) have also been synthesized and characterized [108].

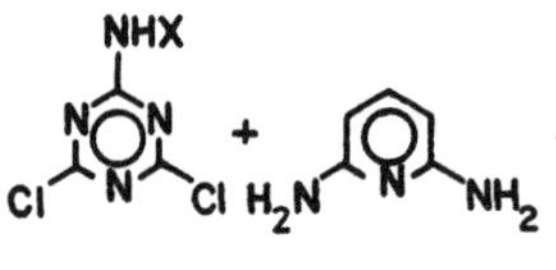

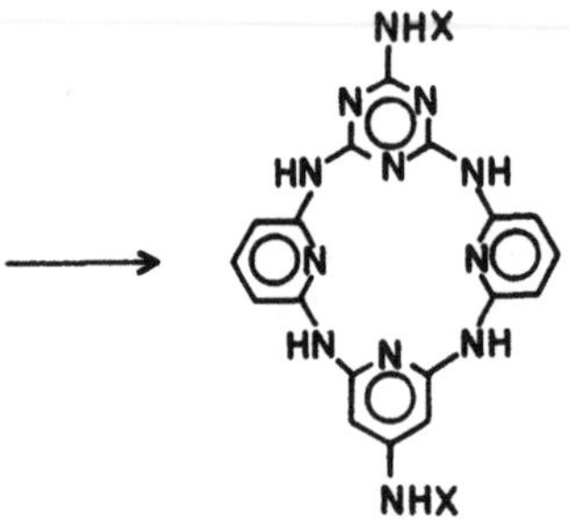

*115* , X=

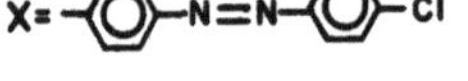

*116* , X=

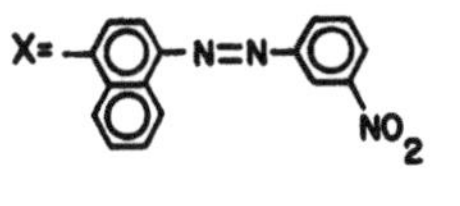

*117* , X=

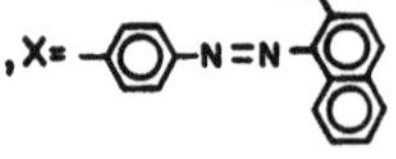

*118* , X=

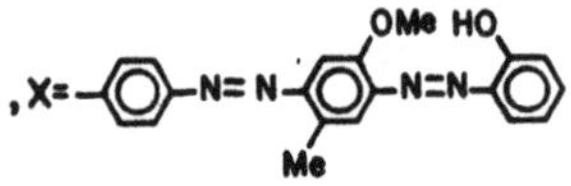

*119* , X=

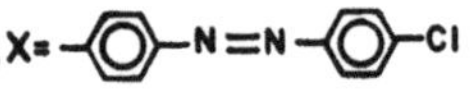

*120* , X=

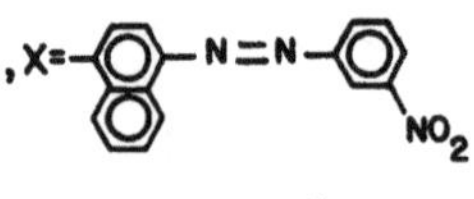

*121* , X=

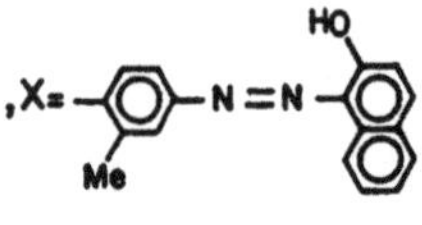

*122* , X=

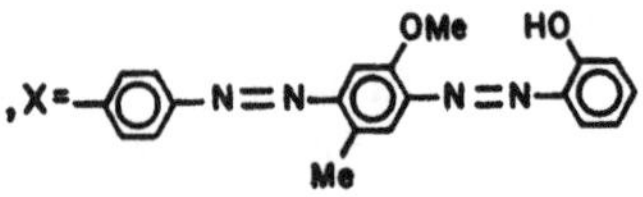

### 3.1.1.3 Syntheses of Coronands Bridged by Hydrazine Moieties

Lewis and Wainwright [109] have prepared a new thirteen-membered macrocycle by treating the nickel(II) or cobalt(II) complexes of 6,6'-dihydrazino-2,2'-dipyridine (*123*) with acetone. In the case of *123* the metal ion chelates very strongly to the dipyridine moiety and only weakly at the terminal nitrogen sites of the hydrazine moiety, rendering these atoms unusually labile and nucleophilic. Treatment of the nickel(II) complex of *123* with refluxing aqueous acetone results within a few minutes in the quantitative formation of the Curtis-type macrocycle (*124*) [109]. These are the mildest conditions yet observed for such a transformation. The metal ion can be displaced with sodium cyanide to afford the free ligand *125*. The iron(II) and cobalt(II) complexes of *125* have also been isolated [109].

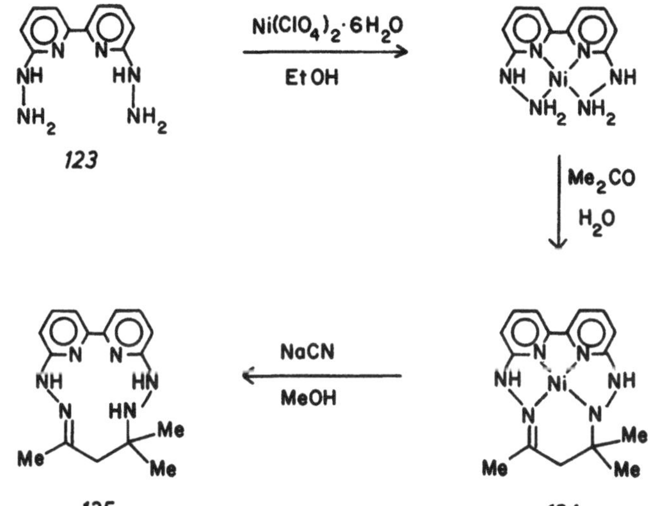


*123* *125* *124*

An analogous system involved the condensation of pentane-2,4-dione (*127*) with 6,6'-*bis*(*N*-methylhydrazino)-2,2'-dipyridine (*126*) in the presence of nickel(II) to generate the thirteen-membered complex *128*. Treatment of the complex with sodium cyanide does not release the ligand, nor does the complex form when copper is used as the metal template [110].

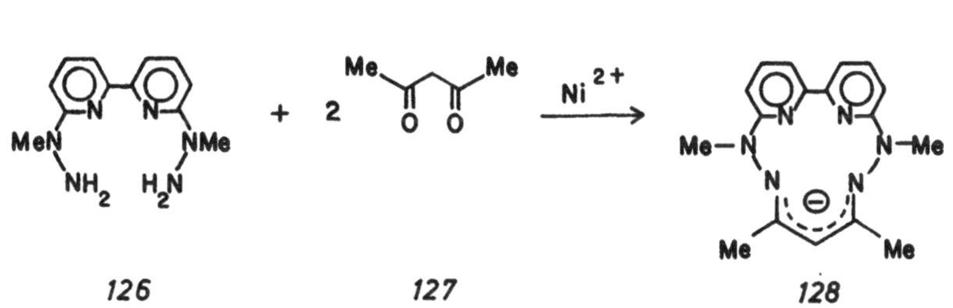


*126* *127* *128*

Treatment of oxovanadium(IV) complexes of 2,6-dipicolinoyl dihydrazine with β-diketones results in the formation of macrocyclic complexes (*129–131*). These products have been characterized on the basis of their elemental analyses, electrical conductance, magnetic susceptibility, and infrared spectral data. [111].

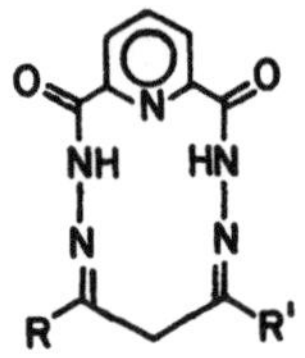

*129* R = Me ; R' = Me

*130* R = Et ; R' = Me

*131* R = Bz ; R' = Me

The only other example of a Schiff base condensation with dipyridine to form a macrocyclic complex has been reported by Tasker et al. [112] in which 2,6-diformylpyridine or 2,6-diacetylpyridine is condensed with 6,6'-dihydrazino-2,2'-dipyridine (*132*) in the presence of zinc(II) to afford a pentagonal bipyramidal complex *133*. The rigidity of the system results in an equatorial '$N_5$' donor set that is essentially planar; water molecules occupy the axial positions.

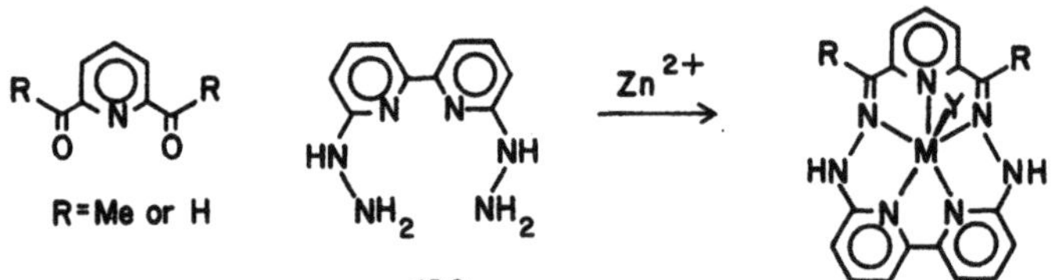

*132*

*133*

R = Me or H

Y = $H_2O$

An iron(II) complex of a completely conjugated fourteen-membered hexa-aza ligand has been prepared by the condensation of 2,6-diacetylpyridine with hydrazine in acetonitrile. There are two possible isomers, *134* and *135*. X-ray data show that the structure of this complex is best represented by *134* in which the macrocycle is completely planar and the coordination geometry is pentagonal bipyramidal. The iron(II) complex of *134* is extremely labile to nucleophiles [113]. The cobalt(II) [113], magnesium(II) [114], zinc(II) [114], and scandium(III) [115] complexes of this hexa-aza quadradentate ligand have also been prepared.

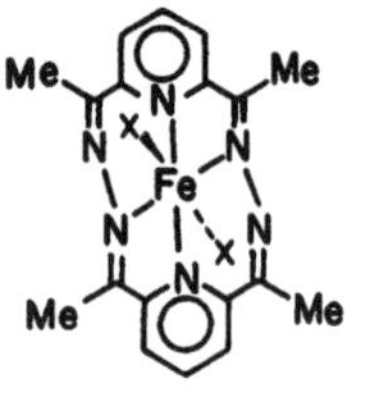

X = $CH_3CN$

*134*

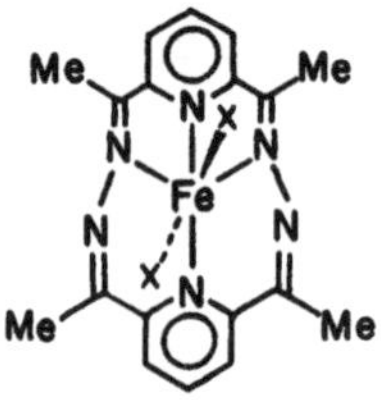

X = $CH_3CN$

*135*

Rana and Teotia [116] have reported the synthesis of a novel tridentate macrocycle *136* by the reaction of 2,6-dipicolinic acid hydrazide and acetylacetone in the presence of cobalt(II), nickel(II), and copper(II) salts. As is frequently the case, water occupies the axial positions of these trigonal bipyramidal complexes and the chelating nitrogen atoms are essentially coplanar.

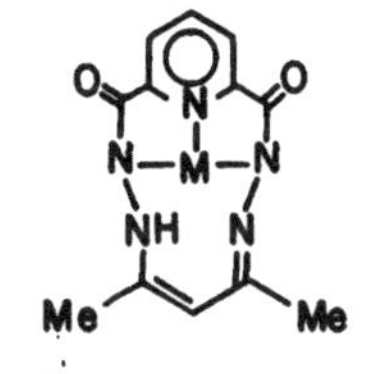

*136*

3.1.2 Synthesis of Coronands by Condensation of Diacid Halides with Polyamines

Vögtle et al. have employed the reaction of 2,6-*bis*(chloroformyl)pyridine with various polyamines in order to synthesize a number of coronands, such as *137–139*. Copper(II) complexes of *137* and *138* have been reported [117].

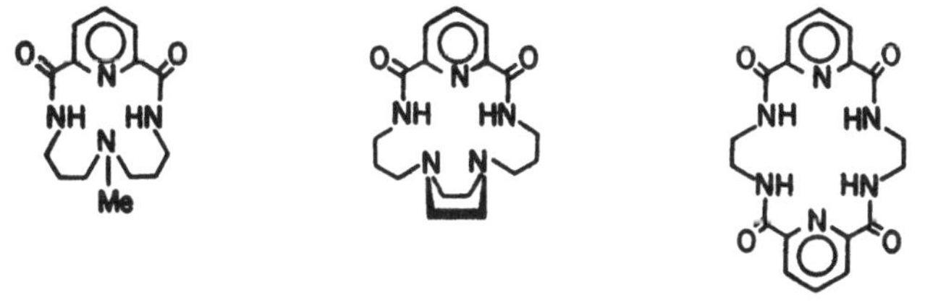

*137* *138* *139*

In 1976, Weber and Vögtle also reported the syntheses of *140*, obtained by treatment of the same substrate with 2,6-diaminopyridine [118]. To date, no complexes of *140* have been reported. Similar macrocycles, prepared by Vögtle, include *141–143* [119], again, no complexes of these ligands have yet been observed.

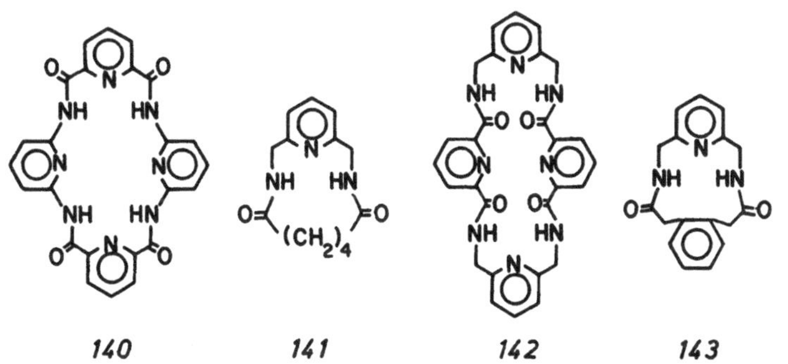

*140* *141* *142* *143*

In view of the affinity of transition metals for dipyridine, this moiety has been incorporated in macrocyclic systems in lieu of pyridine. Vögtle has utilized this new entity in the synthesis of *144* (41 %) and *145* (19 %), but to date no complexes of these two ligands have been isolated.

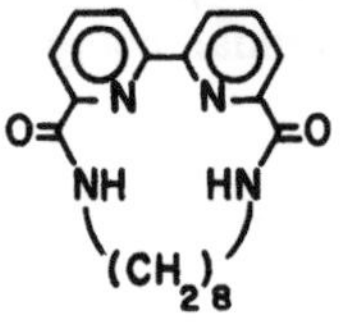


*144*

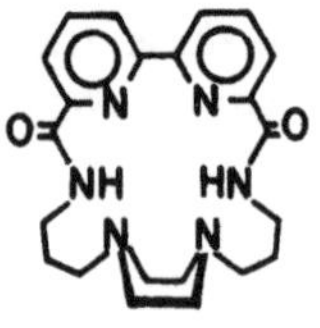


*145*

However, several intriguing macrocyclic ligands which form dinuclear complexes have been synthesized by Lehn and coworkers. Ligands *146* and *147* are synthesized by a route which results in the formation of uncomplexed ligands (Scheme IX). Both ligands form dinuclear copper(II) complexes, in which the copper-copper distance in *146* is 4.79 Å [120].

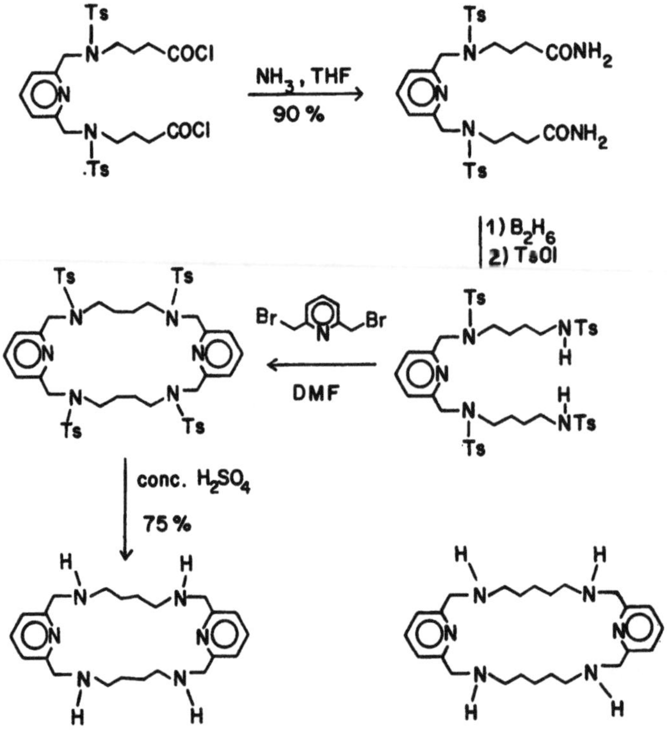


*146* *147*

*Scheme IX*

### 3.1.3 Syntheses by the Nucleophilic Substitution of Dihalides by Diamines

The reaction of 2,6-dichloropyridine with *sym*-dimethylethylenediamine utilizing butyllithium or lithium hydride as the base afforded dimer *148* (1.8 %) and trimer *149* (2 %) [121]. In the $^{1}$H NMR spectrum of *148* at 30 °C, the methylene protons are coalesced. Therefore, variable temperature NMR data afford a calculated $\Delta G^{\ddagger}$ at $T_c = 300$ K of 12.5 kcal/mole, whereas the VT NMR data for *148* can be compared to that of *150* which has a very similar $\Delta G^{\ddagger}$ value of 13.5 kcal/mole. Explained in terms of "anti, longitudinal" isomerization, this isomerization proceeds through a "stepped" intermediate or transition state (Scheme *X*). This conformational motion is believed to be due to the nearly coplanar geometry imposed by the imidate groups in addition to the symmetry considerations [122]. The conformational motion as seen for *148*, according to the VT NMR studies, is also believed to be of the "anti, longitudinal" type, due to the coplanar geometry imposed in this case by the amidine moieties which also have a low dihedral angle ($<30°$) [122–129].

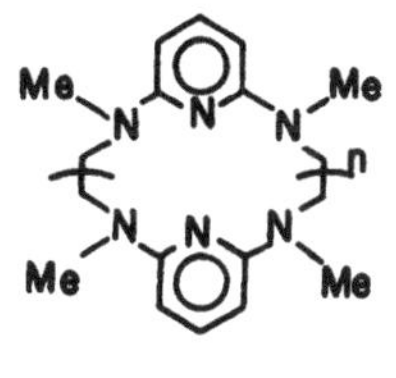

*148* n=1

*149* n=2

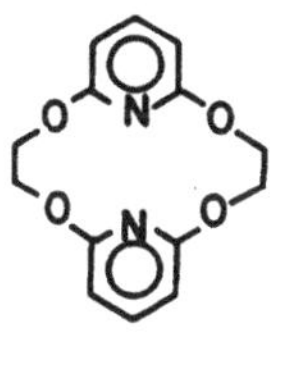

*150*

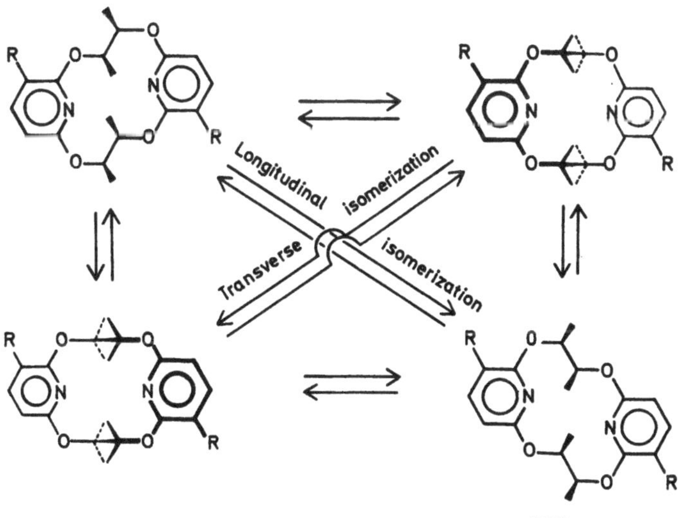


*150*

*Scheme X*

Hexaza "18-crown-6" *151* (7.8%) and *152* (8.0%) have also been synthesized by treatment of 2,6-*bis*(chloromethyl)pyridine with *sym*-dimethylethylenediamine or piperazine in dimethylformamide, utilizing potassium carbonate as the base. The cobalt(II) and copper(II) complexes *153* of *151* have been made and the single crystal X-ray structure determination of each shows that in both cases the metal ion is octahedrally coordinated by the six nitrogen atoms of the ligand. In the case of ligand *152* only the copper(II) complex has been formed, however, at this point the X-ray crystal structure of this complex has not been completely elucidated [121].

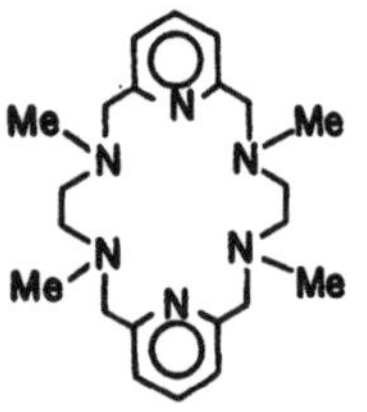


*151*

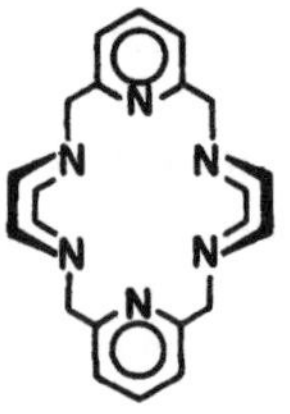


*152*

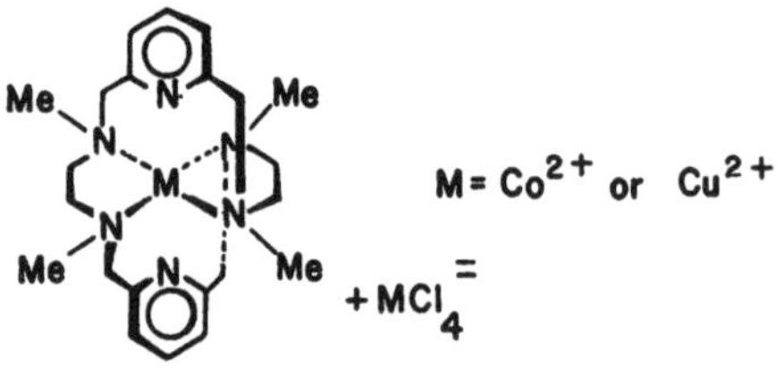


*153*

Most of the macrocycles reported herein have been synthesized by a Schiff base condensation with only a few compounds having been synthesized by alternate methods. Therefore, the major thrust of research in the future should be in devising new methods for synthesizing novel macrocyclic systems and in continuing the study of the effects of structural variations on a macrocycles ability to complex metal ions.

## 3.2 Coronands Containing Oxygen and Nitrogen Heteroatoms

As with the aza coronands previously discussed, the majority of the coronands containing oxygen and nitrogen have been synthesized by Schiff base condensation or *via* a nucleophilic substitution of a halide on a compound processing either pyridine or dipyridine subunits.

### 3.2.1 Synthesis by Schiff Base Condensation

Analogous to the preparation of strictly nitrogen containing macrocycles, *154* [59,78,79,83,87], *155* [83,87], *156* [130,131], and *157* [130,131] have been synthesized by condensation of 2,6-diacetylpyridine or 2,6-diformylpyridine with the desired diaminopolyether in the presence of a metal ion. The metal ions which have been utilized in the synthesis of *154* are iron(III) [78,83], manganese(II) [59,78,83], zinc(II) [59,78,83], magnesium(II) [59,78,132], iron(II) [59,78], cobalt(II) [132], and cadmium [59]. In the case of *155*, only iron(III) [83], manganese(II) [83], zinc(II) [83], and magnesium(II) [132] have been employed. Complexes of *156* and *157* have also been prepared with barium [87,133], strontium [87], calcium [87,133], and lead(II) [130]. The crystal structure determination

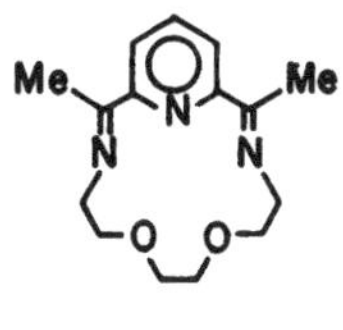

*154*

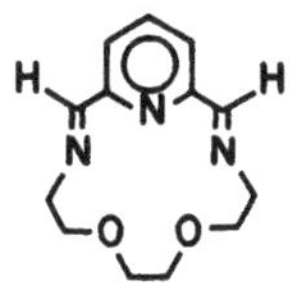

*155*

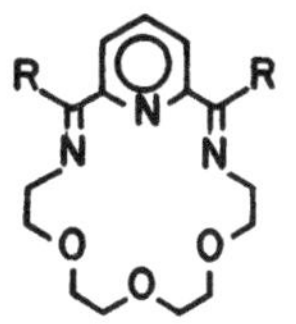

*156* R=H

*157* R=$CH_3$

of the lead complex of *156* indicates that the lead ion is located within the macrocyclic cavity and appears to interact preferentially with the nitrogen atoms [130].

Manganese(II) and zinc(II) metal ions have been used to synthesize *158* and *159*. The coordination geometry of these complexes is distorted pentagonal bipyramidal [97,131]. If manganese(II) chloride in ethane is utilized, *158* is isolated. However, if magnesium(II) perchlorate or magnesium(II) isothiocyanate in alcohol is employed, *159* is isolated. Because both ligands can be formed even in the absence of metal ions, there arise several questions about the factors that control the cyclization. The metal ion may indeed act as a template, or the macrocycle formation may be governed strictly by thermodynamic considerations [131].

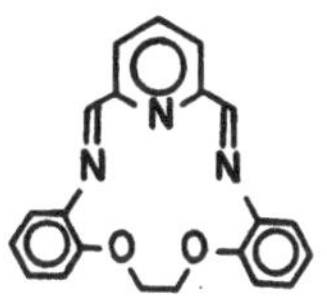

*158*

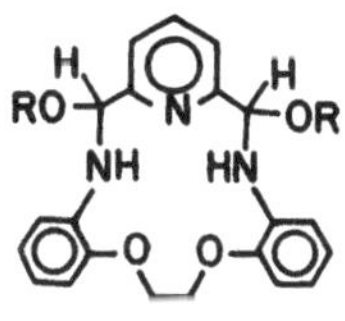

*159*

R=Et, Me, n-Bu

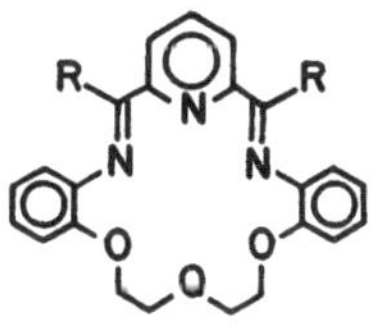

*160* R=H

*161* R=$CH_3$

The calcium, strontium, barium, and lead [80] complexes of *160* and *161* have also been reported. In these two ligands the six donor atoms are essentially confined in a plane; these complexes thus permit study of unusual coordination geometries in species of high coordination number. Attempts to form alkali metal complexes with *160* and *161* under the same conditions as employed for the alkaline earth metal complexes have failed. The successful syntheses of complexes of the latter type indicate that the higher charge to radius ratio is of consequence when spherically charged cations are employed. Such metal ions have no apparent coordinative discrimination as the template ion [87].

A bimetallic complex of a thirty-membered decadentate macrocyclic ligand *162* was synthesized in greater than 80% yield when lead(II) thiocyanate was used as the template. No monomeric complex has yet been isolated from the lead(II). The x-ray crystal structure shows that the coordination geometry of the metal is basically hexagonal bipyramidal with one equatorial site unoccupied. The axial positions are occupied by thiocyanate molecules. Because the metal ions are associated more strongly with the nitrogen atoms than with the oxygen atoms indicates that the weakly coordinating ethereal oxygens may not be able to compete effectively with solvent molecules for coordination sites. As a result, the terminal amine and carbonyl functions are not constrained into close proximity, and there is a preferential self-condensation to afford the dimeric dinuclear complex [80,89].

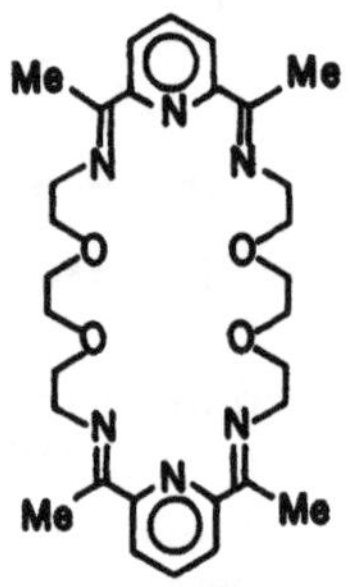

*162*

The dinuclear copper(II) complex of ligand *162* has also been isolated by transmetallation. This copper complex adopts a folded conformation which allows for intramolecular bridging of the metal ions by small anions such as azide and hydroxide. The structure of the μ-azide complex has been determined by x-ray crystal analysis and shows that each copper atom is bound only to the nitrogen atoms of the macrocycle and an azide molecule bridging the two metal ions. There is also a terminal azide ion bound to each copper atom in the plane of the pyridine ring [134]. The transmetallation of complex *162* has also been successfully utilized for iron(II), cobalt(II), cobalt(III), and nickel(II). On the basis of preliminary X-ray data, all of these complexes are believed to be octahedral, the nitrogen atoms being coordinated to the metal ion [135].

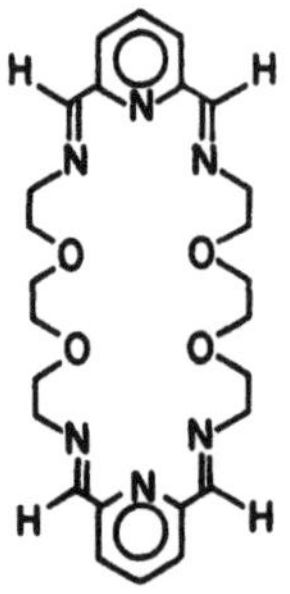

*163*

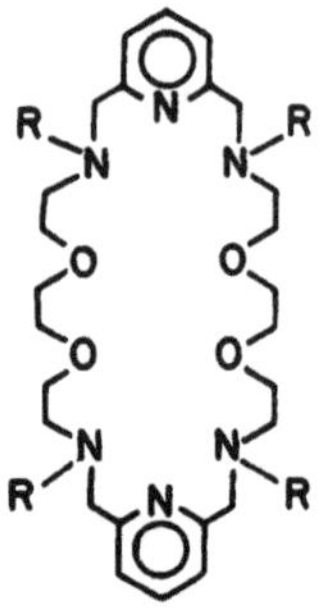

*164* R=H

*165* R=$CH_3$

Ligand *164* has been obtained by the sodium borohydride reduction of *163*, whereas *165* is formed by *N*-methylation of *164*. Copper(I) complexes of *164* and *165* rapidly absorbs dioxygen in equimolar amounts; the ligand than undergoes a slower anaerobic oxidative dehydrogenation to regenerate the copper(I) centers so that the cycle may be repeated at least once again, albeit more slowly. Evidence indicates that the $—CH_2—NH—$ bond is initially oxidized to the imine, then the second, slower dehydrogenation is believed to be that of $—CH_2—CH_2—$, which has a higher activation energy for dehydrogenation than do the $—CH_2—NH—$ groups. Attempts to isolate pure samples of the oxidized complex have proven fruitless [136].

### 3.2.2 Synthesis by Condensation of a Diacyl Halide and a Diaminoether

Another method employed for the synthesis of these types of macrocycles involves the reaction of a *bis*-acid chloride with a diaminoether. Weber and Vögtle have reported the synthesis of a series of these macrocycles and their complexes. The first examples included ligands *166–171* [137], which were obtained in yields varying from 3% to 61%; no metal complexes have as yet been reported for these lactam-type macrocycles [117,137].

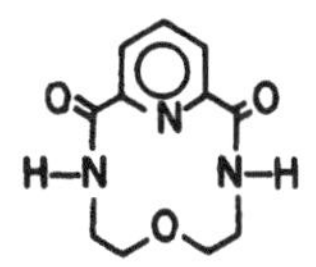

*166* n=1
*167* n=2
*168* n=3

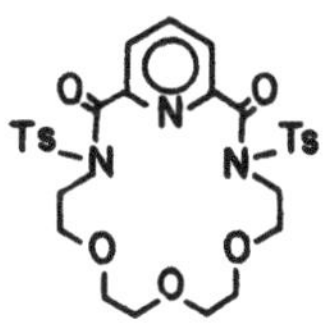

*169*

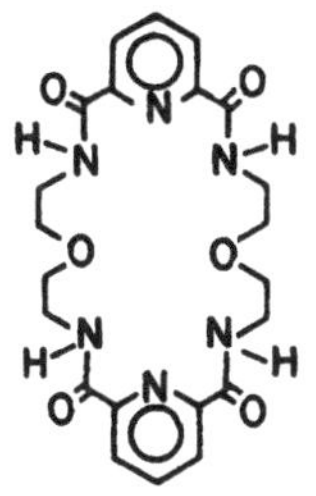

*170*

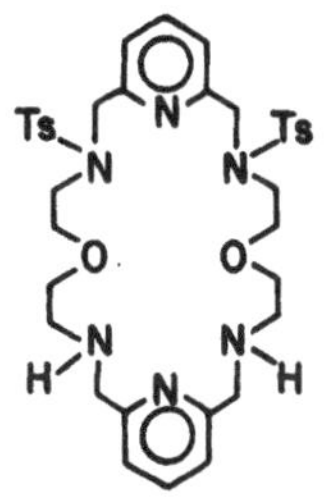

*171*

In 1978, Vögtle reported the synthesis of a different series of lactam-type macrocycles, which were achieved by the reaction of 2,6-*bis*(aminomethyl)pyridine with various polyether-*bis*-acid chlorides [138]. Among the coronands reported were *172* and *173*; to date, only the magnesium complex of *173* has been prepared.

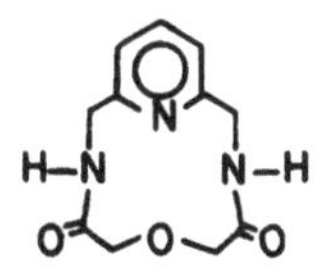

*172*

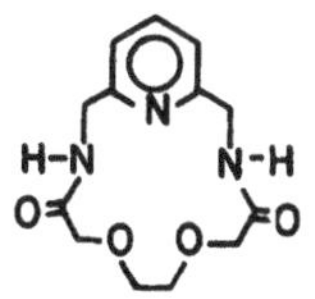

*173*

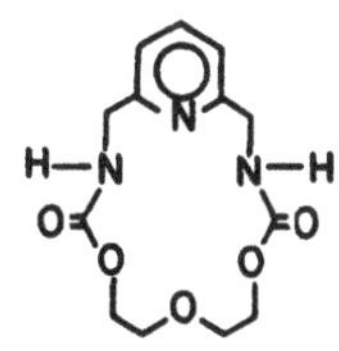

*174*

A slight modification of reactants has led to the synthesis of *174*. In this case, 2,6-*bis*(aminomethyl)pyridine was treated with *bis*(chloroformyl)diethyleneglycol ether to afford *174* (66%); to date there have been no complexes reported for this ligand [138].

This class of macrocycles has recently been expanded to include the 2,2′-dipyridine subunit within the macrocyclic framework. Vögtle et al. have synthesized *175* (59%) and *176* (24%). Thus far, no complexes of these two ligands have been reported [119].

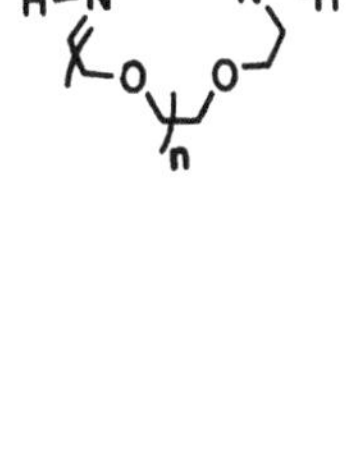

*175* n=1
*176* n=2

### 3.2.3 Synthesis by Nucleophilic Substitution of a Dihalide by an Amino Glycol

Macrocycle *177* in which the 2,6-pyridino and the 1,4-piperazino moieties were incorporated into the macrocyclic framework has been reported. The synthesis of *177* (10%) was accomplished by treatment of 2,6-dichloropyridine with the dianion of *N,N'-bis*(2-hydroxyethyl)piperazine in refluxing xylene. Attempts to prepare the cobalt(II) complex of *177* resulted in diprotonation of the macrocycle. The X-ray crystal structure determination analysis has been performed for both *177* and *178*. According to the crystal structure analysis of *178*, the piperazine rings are in the "chair" conformation in the solid state and the molecule is fairly rigid due to the imposed steric constraints of the imidate moieties [139].

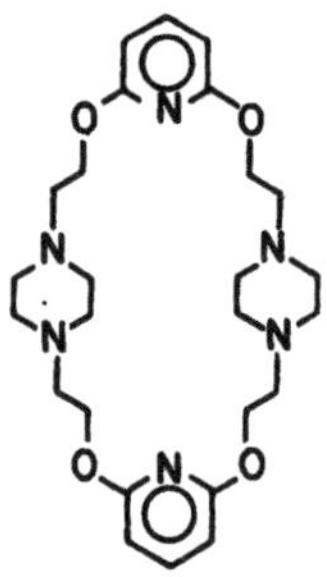


*177*

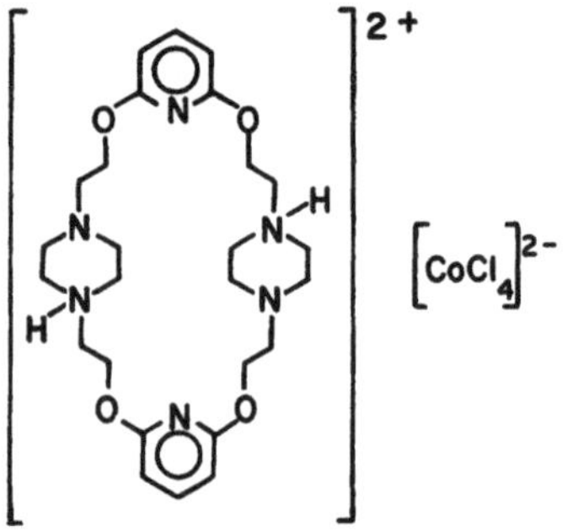


*178*

Ligands *179* and *180* were synthesized by the nucleophilic substitution of the sodium glycolate of *N*-methyldiethanol amine on either 2,6-dichloropyridine or 2,6-*bis*(chloromethyl)pyridine. However, *183* and *184* were synthesized by the quaternization of *181* or *182* with 1,2-*bis*(β-ethoxy)ethane in acetonitrile. In both instances the resulting diquaternary ammonium salts were demethylated by L-Selectride® in refluxing tetrahydrofuran to afford the desired pyridino coronand. Complexation studies have not been performed on any of these coronands and the physical properties of these compounds do not indicate any unusual characteristics [121].

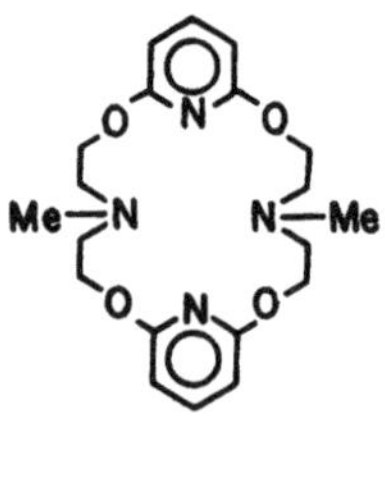


*179*

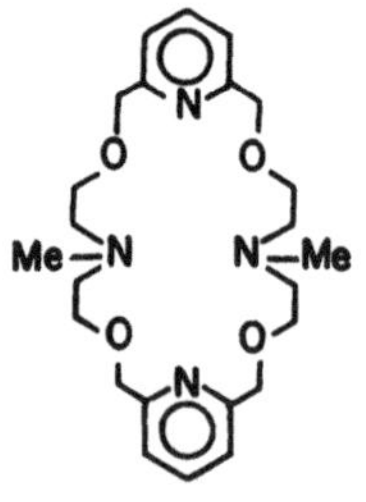


*180*

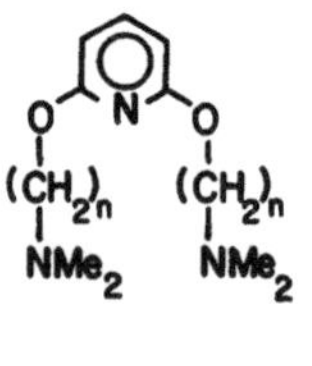


*181* n = 2

*182* n = 3

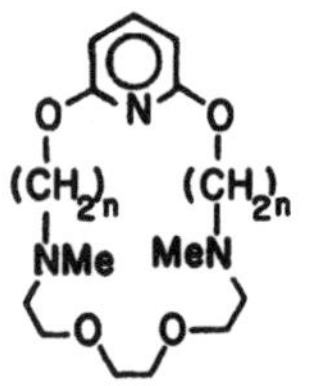


*183* n = 2

*184* n = 3

This discussion has covered all the coronands synthesized to date which contain nitrogen and oxygen heteroatoms or strictly nitrogen atoms in addition to a pyridino subunit in the macrocyclic framework. It has also included a discussion of any complexes which may have been reported. Much of the work has been centered about two synthetic strategies; the amide reduction and the formation *via* Schiff bases. In addition, most of the complexes formed have been with Schiff bases involving transition metals or with the amide type macrocycles utilizing alkali metals or the alkaline earth metals. Therefore, it is more than evident from the information presented herein that much work remains to be done on determining the general structure of these two dimensional ligands along with studying their complexation properties.

# 4 Pyridinocryptands

## 4.1 Synthesis by Condensation of Diacyl Halides and Diamines

Very few of the cryptands which have been synthesized contain the pyridino subunit, notable exception being ligands *185–189*. Reported by Wehner and Vögtle in 1976, these cryptands were synthesized by the condensation of pyridine-2,6-*bis*(carboxylic acid chloride) with the appropriate macrocyclic diazo crown ether. Cryptand *189* was the only amide reduced with diborane to afford the corresponding diamino ligand *186*, which was found to form complexes with $LiClO_4$, NaSCN, and KSCN. Ligand *185* complexed with $LiClO_4$ only. However, cation selectivity and stability measurements were made for several cations; the order of stability for both of the ligands is: $Ba^{2+} > Sr^{2+} > Ca^{2+} > Na^{+} > K^{+} > Li^{+} > Rb^{+} > Cs^{2+} \sim Mg^{2+}$ [140].

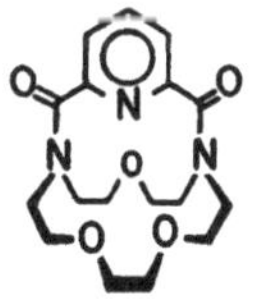

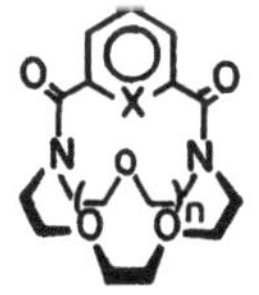

*185*

*186* X=N (63%)
*187* X=CH

*188* X=N; n=1 (29%)
*189* X=N; n=2 (74%)
*190* X=CH

Vögtle has also reported the synthesis of an analogous system (*191* and *192*) which possess a dipyridine bridging unit. The only complexes of ligand *191* reported thus far are of NaSCN, $LiClO_4$, and $NaClO_4$ [141].

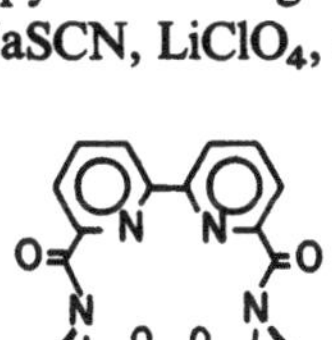

*191*

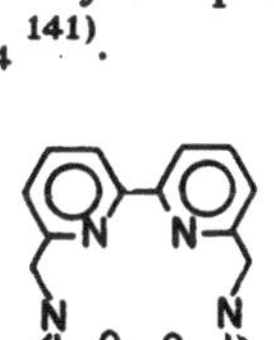

*192*

Lehn [120] has also reported the synthesis of a lateral macrobicyclic molecule, *193*, where a macrocyclic unit is bridged by a chelating unit which contains a

2,6-pyridino subunit. This was accomplished by linking the 2,6-*bis*(aminomethyl)-pyridine unit to a diaza-dithia-12-membered macrocycle by two butane bridges. Such complexes may be referred to as lateral macrobicyclic dinuclear cryptates. These cryptands are of interest due to their ability to form dinuclear complexes with one cation complexed by the macrocycle and a second cation complexed by the bridging subunit. Dinuclear complexes are of particular interest for: the study of cation interaction at short distances; as bio-inorganic models of the metalloproteins when larger interaction distances permit insertion of a substrate molecule; as a means of complexing two different cations, allowing for the possible stabilization of different oxidation states; for the selective fixation and transport of substances (i.e., gases); and for the catalysis of multicenter-multielectronic processes.

Cryptand *193* forms a cryptate containing two copper(II) ions. The reduction potentials of the two ions would be expected to be vastly different, because the $Cu^{2+}$ complex of *194* is reduced at a potential 500 mV less positive than the $Cu^{2+}$ complex of *195*. This is indeed found to be the case. The dinuclear $Cu^{2+}$ complex of *193* undergoes a one-electron reduction at +550 mV. The second $Cu^{2+}$ ion is reduced at +70 mV. Therefore, the first reduction must be that of the $Cu^{2+}$ ion complexed by the [12]-$N_2S_2$ subunit, resulting in the facile formation of a Cu(I)-Cu(II) mixed valence dinuclear cryptate (*196*). This system suggests the possibility of the formation of heterometallic dinuclear complexes [120].

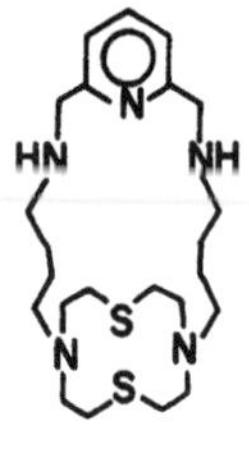


*193*

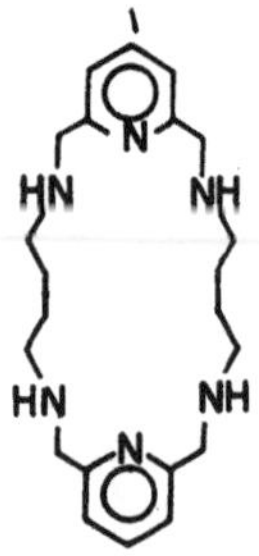


*194*

*195*

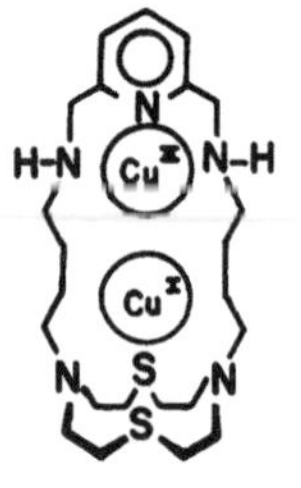


*196*

Gunter and coworkers have synthesized a lateral macrobicyclic molecule *197* where a phorphyrin ring is capped by a bridging unit which contains a pyridino subunit. This model system for cytochrome *c* oxidase is in fact a heterodinuclear complex with iron(III) complexed to the porphyrin unit and copper(II) complexed to the lateral pyridino bridge. A series of these complexes have been made with

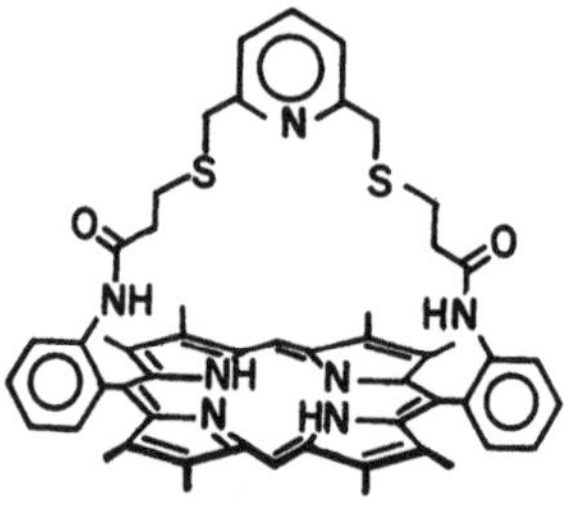


*197*

varying counterions. Mössbauer, ESR, and magnetic moment studies have been conducted to determine if there is any coupling between the two complexed metal atoms, and the data indicate that exchange coupling effects are minimal for the complexes studied [142].

## 4.2 Synthesis by Nucleophilic Substitution of Dihalides with Glycolates

Several cryptands have been synthesized by direct nucleophilic substitution of a halogen. 2,6-Dihalopyridines are subject to nucleophilic substitution by alkoxides, the result being direct attachment of the oxygen atom to the heteroaromatic ring [122, 143]. Cryptand *198* (3%) was formed directly by the reaction of 2,6-dichloropyridine with triethanolamine using sodium hydride as the base for generating the glycolate. The X-ray structure determination demonstrates the symmetrical nature of this molecule which possesses $D_3$ symmetry, deviating from ideal $D_{3h}$ symmetry by a slight twist about the $C_3$ axis [144].

The most surprising aspect of this crystal structure is that the bridgehead nitrogen atoms possess a planar configuration with crystallographically equivalent 120° bond angles. Data indicate that skeletal rigidity is imposed in the molecule by the imidate entity which is associated with a low dihedral angle of 0° (+10°) [122]. Synthetic [145] and theoretical [146] studies have shown that heteroatoms adjacent to the pyridine moiety retard metal ion complexation due to steric problems caused by the preferred conformation of the integrated imidate units and the reduced *N*-electron densities on the pyridine nitrogen atoms [147].

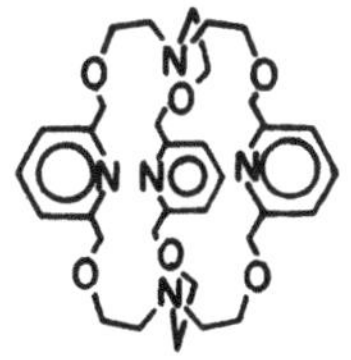

*199*

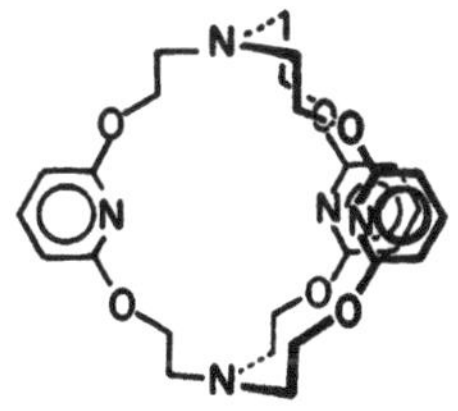

*198*

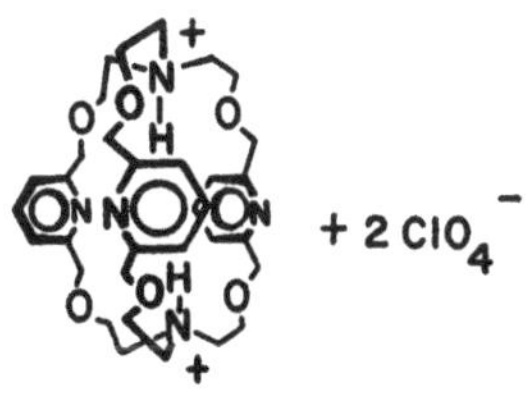

*200*

Due to an interest in studying the effects of structural constraints of the complexing ability of macrocyclic ligands [148], the synthesis of *199*, a structural analog of *198*, in which a methylene group was inserted between each pyridine ring and oxygen atom was accomplished [149]. It was expected that the increased flexibility of the molecule and the removal of the steric constraints caused by the imidate moieties (in *198*) would allow the bridgehead *N*-lone pairs of electrons to assume a nonplanar ($sp^3$) configuration and to thus be available for internal complexation. Cryptand *199* was synthesized by the reaction of 2,6-*bis*(chloromethyl)pyridine and triethanol amine in *N,N*-dimethylformamide with sodium hydride as the base. $^1$H NMR data as well as the X-ray structural determination indicate that one of the pyridine rings is always located within the cavity of the molecule. The low symmetry of *199* and the presence of different conformations for each of its three bridging chains attests to its greater flexibility with respect to cryptand *198*.

The copper(II) chloride and cobalt(II) chloride complexes of *199* were prepared; elemental analyses indicate that in both cases there are two metal associated with each cryptand. To date, however, no crystals suitable for X-ray structural determination studies have been obtained [149].

The diprotonated perchlorate salt *200* of *199* has also been prepared and the X-ray crystal structure determination shows the bridgehead nitrogen atoms to be in the *endo-endo* orientation with the hydrogen atoms on the bridgehead nitrogens located inside the cavity of the molecule. Both of these protons are held in a tetrahedral array by a nitrogen atom and three oxygen atoms. In addition, the three oxygen atoms have turned such that their electron density is located *inside* the cavity in contrast to *199* where the oxygen atoms are oriented such that their electron density is *outside* the cavity [121].

## 4.3 Synthesis by Quaternization-Demethylation

Quaternary ammonium macrocycles [150,151] and cryptands [151] have been synthesized by an alkylation procedure, but no reports of dealkylation and/or *N*-bridge manipulation, other than ring cleavage are known. However, the construction of cryptands has been accomplished from *bis*-diamines by a stepwise quaternization-demethylation sequence (Scheme XI).

Initially, the diquaternary ammonium macrocycle is synthesized by the quaternization of a diamino compound by a diiodoether; the diquaternary salt is, subsequently, demethylated with L-Selectride®. Repeating this sequence of events results in the formation of a cryptand. In most cases the *bis*-quaternary ammonium salts were not

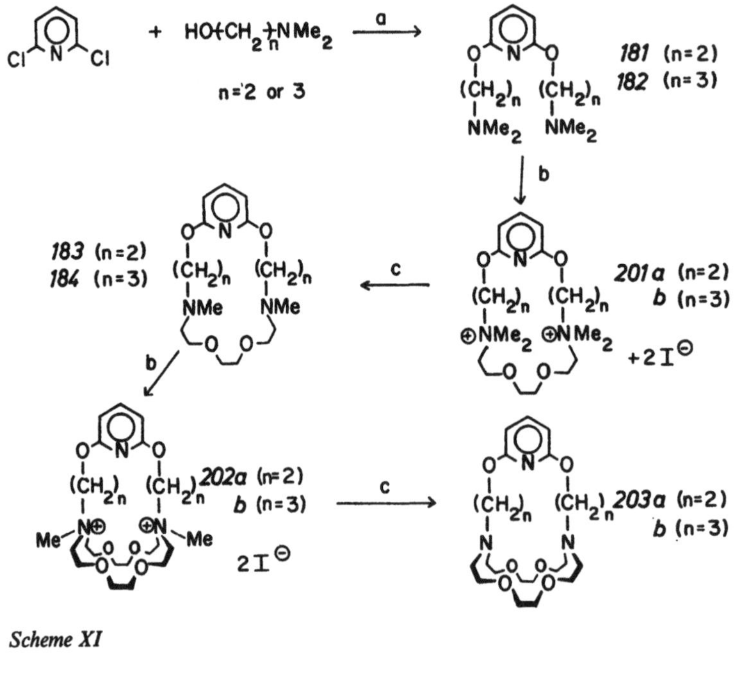


Scheme XI

isolated but were directly demethylated to afford the desired azacrown macrocycle or cryptand.

The single crystal X-ray analysis of *202a*, a *bis*-quaternary ammonium salt, was performed and the compound was shown to be in the anticipated *exo-exo* orientation. No complexation studies have been performed on either the macrocycles or the cryptands [152].

It is more then evident from the data presented herein that although great strides have been made in studying ligand design versus complexation properties the surface has only been scratched. The diverse architecture that characterizes these systems offers the opportunity for significant control over the complexing abilities of these macrocycles. The design and synthesis of new and varied cryptands and studies of their complexation properties offers a unique challenge to chemists, the ultimate goal being the synthesis of molecules which would rival biological systems in their complexation selectivities.

## 5 Acknowledgements

We wish to thank the National Science Foundation, the National Institutes of Health, and Dow Chemical USA for financial support. We also wish to express our sincere appreciation to Hellen Rowland Taylor for editorial comments.

## 6 References

1. Baker, W. et al.: J. Chem. Soc. *1958*, 3594
2. Pahor, N. B., Calligaris, M., Randaccio, L.: J. C. S. Perkin II *1978*, 38
3. Gault, I., Price, B. J., Sutherland, I. O.: Chem. Comm. *1967*, 540
4. Kauffmann, T. et al.: Angew. Chem. Int. Ed. *9*, 808 (1970)
5. Martel, H. J. J-B., Rasmussen, M.: Tetrahedron Lett. *1971*, 3843
6. Bernardi, F. et al.: Chem. Phys. Letts. *36*, 539 (1975)
7. Fletcher, J. R., Sutherland, I. O.: Chem. Comm. *1969*, 1504
8. Bruhin, J., Kneubühler, W., Jenny, W.: Chimia *27*, 277 (1973)
9. Bruhin, J., Jenny, W.: Tetrahedron Lett. *1973*, 1215
10. Vögtle, F.: ibid *1968*, 3623
11. Jenny, W., Holzrichter, H.: Chimia *23*, 158 (1969)
12. Jenny, W., Holzrichter, H.: ibid. *22*, 306 (1968)
13. Boekelheide, V., Lawson, J. A.: Chem. Comm. *1970*, 1558
14. Taylor, D.: Aust. J. Chem. *31*, 1953 (1978)
15. Gompper, R., Seybold, G.: Angew. Chem. Int. Ed. *7*, 824 (1968)
16. Martel, H., McMahon, S., Rasmussen, M.: Aust. J. Chem. *32*, 1241 (1979)
17. Cooke, Jr. M. P.: J. Org. Chem. *46*, 1747 (1981)
18. Tsuge, O., Okumura, M.: Heterocycles *6*, 5 (1977)
19. Haenel, M. W.: Tetrahedron Lett. *1978*, 4007
20. Newkome, G. R., Roper, J. M., Robinson, J. M.: J. Org. Chem. *45*, 4380 (1980)
21. Newkome, G. R., Kohli, D. K.: Heterocycles *15*, 739 (1981)
22. Boekelheide, V., Galuszko, K., Szeto, K. S.: J. Amer. Chem. Soc. *96*, 1578 (1974)
23. Boekelheide, V., Reingold, I. D., Tuttle, M.: J. C. S. Chem. Comm. 1973, 406
24. Weaver, L. H., Matthews, B. W.: J. Amer. Chem. Soc. *96*, 1581 (1974)
25. Lord, R. C., Nolin, B., Stidham, H. D.: ibid. *77*, 1365 (1955)
26. Richman, J. E., Simmons, H. E.: Tetrahedron *30*, 1769 (1974)

27. Reingold, I. D., Schmidt, W., Boekelheide, V.: J. Amer. Chem. Soc. *101*, 2121 (1979)
28. Jenny, W., Holzrichter, H.: Chimia *21*, 509 (1967)
29. Jenny, W., Holzrichter, H.: ibid. *22*, 139 (1968)
30. Deuchert, K., Hünig, S.: Stud. Org. Chem. (Amsterdam) *3*, 202 (1979)
31. Boekelheide, V., Pepperdine, W.: J. Amer. Chem. Soc. *92*, 3684 (1970)
32. Wong, C., Paudler, W. W.: J. Org. Chem. *39*, 2570 (1974); Atwood, J. L. et al., J. Heterocycl. Chem., *12*, 433 (1975)
33. Bruhin, J., Jenny, W.: Chimia *25*, 238 (1971)
34. Bruhin, J., Jenny, W.: ibid *25*, 308 (1971)
35. Bruhin, J., Jenny, W.: ibid. *26*, 420 (1972)
36. Iwata, M., Kuzuhara, H., Emoto, S.: Chem. Lett. *1976*, 983
37. Hanson, A. W.: Cryst. Struct. Comm. *10*, 751 (1981)
38. Hanson, A. W.: ibid. *10*, 313 (1981)
39. Kang, H. C., Boekelheide, V.: Angew. Chem. Int. Ed. *20*, 571 (1981)
40. Hanson, A. W.: Acta Cryst. *B33*, 2657 (1977)
41. Melson, G. A.: Coordination Chemistry of Macrocyclic Compounds (Melson, G. A. Ed.), Chapters 1 and 2. Plenum Press, New York 1979
42. Busch, D. H.: Rec. Chem. Prog. 25, 107 (1964)
43. Black, D. St. C., Markham, E.: Rev. Pure and Appl. Chem. *15*, 109 (1965)
44. Busch, D. H.: Helv. Chim. Acta, Fasciculus Extraordinarius, Alfred Werner, 174 (1967)
45. Lindoy, L. F., Busch, D. H.: Prep. Inorg. React. 6, 1 (1971)
46. Black, D. St. C., Hartshor, A. H.: Coord. Chem. Rev. *9*, 219 (1972–1973)
47. Curry, J. D., Busch, D. H.: J. Amer. Chem. Soc. *86*, 592 (1964)
48. Drew, M. G. B. et al.: J. C. S. Dalton *1976*, 1388
49. Nelson, S. M., Bryan, P., Busch, D. H.: Chem. Comm. *1966*, 641
50. Rich, R. L., Stucky, G. L.: Inorg. Nucl. Chem. Letters *1*, 61 (1965)
51. Karn, J. L., Busch, D. H.: Inorg. Chem. *8*, 1149 (1969)
52. Nelson, S. M., Busch, D. H.: ibid. 8, 1859 (1969)
53. Rakowski, M. C., Rycheck, M., Busch, D. H.: ibid. *14*, 1194 (1975)
54. Fleischer, E., Hawkinson, S.: J. Amer. Chem. Soc. *89*, 720 (1967)
55. Drew, M. G. B. et al.: Acta Cryst. *B32*, 1029 (1976)
56. Drew, M. G. B., Othman, A. H. b., Nelson, S. M.: J. C. S. Dalton *1976*, 1394
57. Drew, M. G. B., Martin, S. M.: Acta Cryst. *A31*, S140 (1975)
58. Drew, M. G. B. et al.: J. C. S. Dalton *1977*, 438
59. Drew, M. G. B. et al.: ibid. *1977*, 1173, and references therein
60. Drew, M. G. B. et al.: J. C. S. Chem. Comm. *1975*, 818
61. Merrell, P. H. et al.: J. Amer. Chem. Soc. *92*, 7590 (1970)
62. Riley, D. P. et al.: Inorg. Chem. *14*, 490 (1975)
63. Lindoy, L. F. et al.: J. Coord. Chem. *1*, 7 (1971)
64. Keypour, H., Stotter, D. A.: Inorg. Chimica Acta *19*, L48 (1976)
65. Barefield, E. K. et al.: Inorg. Chem. *11*, 283 (1972)
65a. Morliere, P., Patterson, L. K.: Inorg. Chem. Acta *64*, L/83 (1982)
66. Long, K. M., Busch, D. H.: ibid. *9*, 505 (1970)
67. Ochiai, E., Busch, D. H.: ibid. *8*, 1474 (1969)
68. Ochiari, E. et al.: J. Amer. Chem. Soc. *91*, 3201 (1969)
69. Farmery, K., Busch, D. H.: Chem. Comm. *1970*, 1091
70. Long, K. M., Busch, D. H.: J. Coord. Chem. *4*, 113 (1974)
71. Ochiai, E.-i., Busch, D. H.: Chem. Comm. *1968*, 905
72. Poon, C.-K., Wan, W.-K., Liao, S. S. T.: J. C. S. Dalton *1977*, 1247
73. Dabrowiak, J. C. et al.: Inorg. Chem. *16*, 540 (1977)
74. Poon, C.-K., Che, C.-M.: J. C. S. Chem. Comm. *1979*, 861
75. Fabbrizzi, L., Micheloni, M., Paoletti, P.: ibid. *1978*, 833
76. Fabbrizzi, L., Micheloni, M., Paoletti, P.: J. C. S. Dalton *1979*, 1581
77. Prince, R. H., Stotter, D. A., Woolley, P. R.: Inorg. Chimica Acta *9*, 51 (1974)
78. Drew, M. G. B., McFall, S. G., Nelson, S. M.: J. C. S. Dalton *1979*, 575
79. Cairns, C. et al.: ibid. *1979*, 446
80. Cook, D. H. et al.: ibid. *1979*, 414

81. Drew, M. G. B. et al.: ibid. *1975*, 2507
82. Nelson, S. M. et al.: J. C. S. Chem. Comm. *1977*, 167
83. Drew, M. G. B. et al.: Inorg. Chimica Acta *12*, L25 (1975)
84. Nelson, S. M. et al.: J. C. S. Chem. Comm. *1977*, 370
85. Black, D. St. C., Rothnie, N. E.: Tetrahedron Lett. *1978*, 2835
86. Drew, M. G. B., Nelson, S. M.: Acta Cryst. *B35*, 1594 (1979)
87. Cook, D. H., Fenton, D. E.: J. C. S. Dalton *1979*, 266
88. Cabral, J. d. O. et al.: Inorg. Chimica Acta *25*, L77 (1977)
89. Drew, M. G. B. et al.: J. C. S. Chem. Comm. *1978*, 415; Drew, M. G. B., Nelson, S. M., Reedijk, J.: Anorg. Chim. Acta *64*, L189 (1982)
90. Drew, M. G. B. et al.: J. Chem. Res., (S), *1979*, 16; (M), *1979*, 0360
91. Stotz, R. W., Stoufer, R. C.: Chem. Comm. *1970*, 1682
92. Backer-Dirks, J. D. J. et al.: J. C. S. Chem. Comm. *1979*, 774
93. Radecka-Paryzek, W.: Inorg. Chimica Acta *45*, L147 (1980)
94. Cabral, J. d. O. et al.: ibid. *30*, L313 (1978)
95. Drew, M. G. B. et al.: J. C. S. Chem. Comm. *1979*, 1033
96. Nelson, S. M. et al.: ibid. *1979*, 1035
97. Alcock, N. W. et al.: ibid. *1974*, 727
98. Lotz, T. J., Kaden, T. A.: ibid. *1977*, 15
99. Lotz, T. J., Kaden, T. A.: Helv. Chim. Acta *61*, 1376 (1978)
100. Sharpless, K. B., Jensen, H. P.: Inorg. Chem. *13*, 2617 (1974); Collman, J. P. et al.: J. Amer. Chem. Soc. *95*, 1656 (1973)
101. Keypour, H., Stotter, D. A.: Inorg. Chimica Acta *33*, L149 (1979)
102. Curtius, N. F.: Coord. Chem. Rev. *3*, 3 (1968)
103. Smirnov, P. R., Gnedina, V. A., Borodkin, V. F.: Tr. Vses. Mezhvuz. Nauch.-Tekh. Konf. Vop. Sin. Primen. Krasitelei., *1970*, 17 [Chem. Abstr. *76*, 14518f (1972)]
104. Borodkin, V. F., Komarov, R. D.: Izv. Vyssh. Ucheb. Zaved., Khim. Khim. Tekhnol. *16*, 1304 (1973) [Chem. Abstr. *79*, 137113q (1973)]
105. Borodkin, V. F., Komarov, R. D.: USSR Pat. 4111,087 (1974) [Chem. Abstr. *80*, 108593m (1974)]
106. Borodkin, V. F., Komarov, R. D.: Izv. Vyssh. Ucheb. Zaved., Khim. Khim. Tekhnol. *16*, 1764 (1973) [Chem. Abstr. *80*, 59924j (1974)]
107. Borodkin, V. F., Komarov, R. D., Aleksandrova, O. A.: Tr. Ivanov. Khim.-Tekhnol. Inst. *1972*, 141 [Chem. Abstr. *79*, 115546f (1973)]
108. Borodkin, V. F., Gnedina, V. A., Grukova, I. A.: Izv. Vyssh. Ucheb. Zaved., Khim. Khim. Tekhnol. *16*, 1722 (1973) [Chem. Abstr. *80*, 70791j (1974)]
109. Lewis, J., Wainwright, K. P.: J. C. S. Chem. Comm. *1974*, 169
110. Lewis, J., Wainwright, K. P.: J. C. S. Dalton *1978*, 440
111. Kher, S. et al.: Inorg. Chimica Acta *37*, 121 (1979)
112. Haque, Z. P. et al.: ibid. *23*, L21 (1977)
113. Goedken, V. L. et al.: J. Amer. Chem. Soc. *96*, 7693 (1974)
114. Radecka-Paryzek, W.: Inorg. Chimica Acta *34*, 5 (1979)
115. Radecka-Paryzek, W.: ibid. *35*, L349 (1974)
116. Rana, V. B., Teotia, M. P.: Ind. J. Chem. *19A*, 267 (1980)
117. Vögtle, F. et al.: Chem. Ztg. *98*, 562 (1974)
118. Weber, E., Vögtle, F.: Ann. Chem. *1976*, 891
119. Bühleier, E., Wehner, W., Vögtle, F.: ibid. *1978*, 537
120. Lehn, J. M.: Pure and Appl. Chem. *52*, 2441 (1980)
121. Majestic, V. K.: Ph. D. Dissertation, Louisiana State Univ., Baton Rouge, LA, 1982
122. Newkome, G. R., Kawato, T.: J. Amer. Chem. Soc. *101*, 7088 (1979); Newkome, G. R. et al.: J. Org. Chem. *44*, 3812 (1979) and references therein
123. Newkome, G. R. et al.: J. Amer. Chem. Soc. *97*, 3232 (1975)
124. Clark, G. R., Palenik, G. J.: ibid. *94*, 4005 (1972)
125. Johnson, J. E., Jacobson, R. A.: Acta Cryst. *B29*, 1669 (1973)
126. Koch, M. H. J. et al.: ibid. *B33*, 1975 (1977)
127. Simon, K., Ibers, J. A.: ibid. *B32*, 2699 (1976)
128. Mente, D. C., Sundberg, R. J., Bryan, R. F.: ibid. *B33*, 3923 (1977)
129. Goldstein, P.: ibid. *B31*, 2086 (1975)

130. Fenton, D. E., Cook, D. H., Nowell, I. W.: J. C. S. Chem. Comm. *1977*, 274
131. Cook, D. H., Fenton, D. E.: Inorg. Chimica Acta *25*, L95 (1977)
132. Cook, D. H. et al.: J. C. S. Dalton *1977*, 446
133. Fenton, D. E. et al.: J. C. S. Chem. Comm. *1978*, 279
134. Drew, M. G. B., McCann, M., Nelson, S. M.: ibid. *1979*, 481
135. Nelson, S. M. et al.: J. C. S. Dalton *1979*, 1477
136. Burnett, M. G. et al.: J. C. S. Chem. Comm. *1980*, 829
137. Weber, E., Vögtle, F.: Chem. Ber. *109* 1803 (1976)
138. Buhleier, E. et al.: Ann. Chem. *1978*, 1586
139. Fronczek, F. R. et al.: J. C. S. Perkins Trans. *1981*, 331
140. Wehner, W., Vögtle, F.: Tetrahedron Lett. *1976*, 2603
141. Buhleier, E., Wehner, W., Vögtle, F.: Chem. Ber. *111*, 200 (1978)
142. Gunter, M. J. et al.: J. Amer. Chem. Soc. *103*, 6784 (1981); Gunter, M. J., Mander, J. N.: J. Org. Chem. *46*, 4792 (1981)
143. Newkome, G. R. et al.: J. Org. Chem. *44*, 3816 (1979)
144. Newkome, G. R. et al.: J. Amer. Chem. Soc. *101*, 1047 (1979)
145. Newkome, G. R. Taylor, H. C. R.: J. Org. Chem. *44*, 1362 (1979)
146. Mattice, W., Newkome, G. R.: J. Amer. Chem. Soc. *101*, 4477 (1979)
147. Newkome, G. R., Kohli, D. K., Fronczek, F. R.: J. C. S. Chem. Comm. *1980*, 9
148. Newkome, G. R. et al.: J. Amer. Chem. Soc. *102*, 7608 (1980)
149. Newkome, G. R., Majestic, V. K., Fronczek, F. R.: Tetrahedron Lett. *1981*, 3035
150. Iabushi, I., Kimura, Y., Yamamura, K.: J. Amer. Chem. Soc. *100*, 1304 (1978); Lagidze, D. R. et al.: Zhur. Organich. Khim. *12*, 2185 (1976); Murakami, Y. et al.: J. C. S. Perkin I *1979*, 1669; Pearson, D. J., Leigh, S. J., Sutherland, I. O.: ibid. *1979*, 3113
151. Schmidtchen, F. P.: Angew. Chem. Int. Ed. Engl. *16*, 720 (1977); Wester, N., Vögtle, F.: Chem. Ber. *113*, 1487 (1980)
152. Newkome, G. R., Majestic, V. K., Fronczek, F. R.: Tehrahedron Lett. *1981*, 3039

# New Perspectives in Polymer-Supported Peptide Synthesis

**V. N. Rajasekharan Pillai**[1] **and Manfred Mutter**

Institut für Organische Chemie, Universität Mainz, D-6500 Mainz, FRG

**Table of Contents**

**1 Introduction** . . . . . . . . . . . . . . . . . . . . . . . . . . 121
1.1 Scope and Organization . . . . . . . . . . . . . . . . . . . . . 121

**2 Polymeric Supports for Heterogeneous Organic Synthesis** . . . . . . . . . 122
2.1 The Solid Phase Peptide Synthesis . . . . . . . . . . . . . . . . . 122
2.1.1 Polystyrene Supports in the Solid Phase Method . . . . . . . . 123
2.1.2 Analytical Control in the Stepwise Synthesis . . . . . . . . . . 125
2.1.3 Mechanization and Automation . . . . . . . . . . . . . . . 126
2.2 Incompatibility of Polystyrene Supports . . . . . . . . . . . . . . 127
2.3 Polyacrylamide-Type Supports for Peptide Synthesis . . . . . . . . . 128
2.4 Poly(*N*-acrylylpyrrolidine) Support. . . . . . . . . . . . . . . . 130
2.5 Segment Condensation on Solid Polymeric Supports . . . . . . . . . 131
2.6 Multipurpose and Multidetachable Polystyrene Supports . . . . . . . 132

**3 The Polymeric Reagent Method** . . . . . . . . . . . . . . . . . . 134
3.1 Polymeric Active Esters in Peptide Synthesis . . . . . . . . . . . . 134
3.2 Racemization in the Polymer Reagent Method . . . . . . . . . . . . 136
3.3 Mechanization of the Polymer Reagent Technique . . . . . . . . . . 136
3.4 Specific Applications, Scope and Limitations . . . . . . . . . . . . 137

**4 Reactivity in Polymeric Solid (Gel) Phase Reactions** . . . . . . . . . . . 138
4.1 Non-equivalence of Reaction Sites . . . . . . . . . . . . . . . . . 139
4.2 Variables of Gel Preparation and Reactivity of Functional Groups . . . 140
4.2.1 Degree of Crosslinking . . . . . . . . . . . . . . . . . . . 140
4.2.2 Topographical Nature of the Polymer Matrix . . . . . . . . . 141
4.3 Intraresin Site Interactions . . . . . . . . . . . . . . . . . . . . 142
4.4 Solvation and Swellability of Peptide Resins. . . . . . . . . . . . . . 145

[1] Permanent Address: Department of Chemistry, University of Calicut, Kerala, 673 635 — India

**5 Techniques with Soluble Polymers** . . . . . . . . . . . . . . . . . . . . . . 147
5.1 Polyethers as Soluble Supports . . . . . . . . . . . . . . . . . . . . 149
5.2 Peptide Synthesis on Polyethyleneglycol Supports . . . . . . . . . . . 151
5.3 The Technique of the Liquid Phase Method. . . . . . . . . . . . . . 152
5.3.1 Attachment of the First Amino Acid and Anchoring Groups. . . 152
5.3.2 Coupling and Deprotection Steps. . . . . . . . . . . . . . . . . 156
5.3.3 Analysis and Automation . . . . . . . . . . . . . . . . . . . 157
5.4 Solid-Liquid Phase Techniques . . . . . . . . . . . . . . . . . . . 158
5.5 Solubilizing Effect of Polyethyleneglycol Supports and Conformational Analysis of Peptides . . . . . . . . . . . . . . . . . . . . . . . . 160
5.6 Conformational Aspects in Peptide Synthesis . . . . . . . . . . . . 164
5.6.1 Chain Elongation and Physicochemical Properties . . . . . . . . 164
5.6.2 Implications in Synthetic Strategies . . . . . . . . . . . . . . . 167

**6 Conclusions and Outlook** . . . . . . . . . . . . . . . . . . . . . . . . 168

**7 Acknowledgements**. . . . . . . . . . . . . . . . . . . . . . . . . . . 169

**8 References** . . . . . . . . . . . . . . . . . . . . . . . . . . . . . 169

Abbreviations used for amino acids and protecting groups throughout this article follow the rules of the IUPAC-IUB Commission on Biochemical Nomenclature; see: J. Biol. Chem. 247, 997 (1972); Int. J. Pept. Protein Res. 7, 91 (1975). In addition, the following abbreviations are also used: PEG, polyethyleneglycol; MPEG, monofunctional polyethyleneglycol; PAP, poly(*N*-acrylylpyrrolidine), DCC, dicyclohexylcarbodiimide; HOBt, 1-hydroxybenzotriazole; DMF, dimethylformamide; TFE, trifluoroethanol; TFA, trifluoroacetic acid; DVB, divinylbenzene; Boc, *N*-*tert*-butyloxy carbonyl.

# 1 Introduction

The solid phase peptide synthesis introduced in 1963 by Merrifield makes use of an insoluble, functionalized polystyrene support meant as an inert carrier onto which the amino acid residues are stepwise incorporated [1]. The operational simplicity of this polymer support procedure for peptide synthesis over the conventional methods has induced intense activity in peptide chemistry during the last two decades. The scope of this method in the synthesis of peptides and even proteins has been demonstrated by the synthesis of a number of biologically active hormones such as Ribonuclease A [2] and human growth hormone [3]. The method has also been applied to the synthesis of oligonucleotides [4] and oligosaccharides [5]. The polymer support method found application in various other fields like general organic synthesis [6], analytical chemistry [7], enzyme immobilization [8], catalysis [9] and mechanistic organic chemistry [10]. The scope and limitations of the solid phase method of peptide synthesis using cross-linked polystyrene supports have been the subject of a number of reviews [11]. Birr in his monograph on the Merrifield's method of peptide synthesis has given a critical evaluation of the chemical details and applicability of this technique [12].

In the stepwise synthesis of biopolymers, particularly of peptides, on cross-linked macromolecular supports it was generally expected that the support would act only as an inert, solid carrier. However, during the last decade innumerable investigations dealing with the quantitative aspects of polymer-supported reactions have shown that the insoluble support does have a significant dynamic influence on the bound substrates. The rates of the reactions, solvation of the bound species, and the mass transport of reagents and solvents in these reactions have been observed to be largely reduced, and this factor, in a way, obstructed a more general application of the solid phase method using cross-linked polystyrenes. Systematic studies of the interdependence of these factors and support characteristics have contributed very much for the improvements in the design of the original Merrifield resins and also for the evolution of a number of related strategies for polymer-bound peptide synthesis.

## 1.1 Scope and Organization

This paper is not intended as a general review covering exhaustively the literature on syntheses of peptides using polymeric supports. Rather, we analyze the studies which helped both the optimization of the polymeric supports for solid phase peptide synthesis and the development of a number of new related strategies for peptide synthesis. First, after stating briefly the salient features of the Merrifield technique, the most significant recent developments in the solid phase synthesis are summarized. These include the introduction of new polar polymeric supports which circumvent the physicochemical incompatibility of the polystyrene resins, solid phase segment condensation, polymer reagent methods and the use of multidetachable resin supports. How the interdependence of the reactivity factors and the solvation, swelling and solubilizing properties of resin supports provides guidance in the selection of suitable polymeric supports for peptide synthesis is illustrated next. Thirdly, the factors which led to the development of soluble polymer techniques for peptide synthesis are delineated; the liquid phase method using hydrophilic polymeric supports is a significant step in this field and the application of this method for the facilitation of conformational studies is also dealt with. The impact of conformational features of the bound peptides on the stepwise and segment condensation approaches for their synthesis forms yet another theme of discussion.

## 2 Polymeric Supports for Heterogeneous Organic Synthesis

Polymers which find application as heterogeneous supports for carrying out organic reactions fall into two main types: the passive supports and the active supports.

In passive supports, the polymer serves as a heterogeneous matrix to which a low molecular weight substrate is
1. covalently bound,
2. allowed to react with various reagents, and
3. cleaved from the support in a modified form;

the solid phase syntheses of peptides, nucleotides and oligosaccharides are typical examples. This concept of passive participants in organic synthetic reactions increases the efficiency of intermediate manipulation and product recovery by attaching soluble substrates to insoluble supports.

In active supports, the substrate attached to the insoluble polymer effects a synthetic or a catalytic transformation on a soluble substrate. Thus, this group includes the *polymer-bound reagents* [13–15], in which the active site is consumed during the course of the reaction, and the *polymer-bound catalysts* [9], in which the active site on the heterogeneous matrix effects catalytic transformations on a soluble substrate.

The immobilized polymeric reagents are similar in behaviour to their low molecular weight analogs, but have the inherent advantage of insolubility. In solvent-swollen polymer reagents, the microenvironment of the reactive site is essentially the same as that encountered by reactants in solution; however, the bulkiness of the insoluble phase imposes steric contraints on the transition state which may lead to regioselective and/or stereospecific transformations. Moreover, the high local concentration of the reagents in the polymer matrix significantly influences the kinetic course of the reaction. The desirable properties of an effective polymer-bound reagent are its filterability, facile access to and egress from its active site by soluble reagents and solvents and regeneratability of the active site for subsequent use [15].

### 2.1 The Solid Phase Peptide Synthesis

In the solid phase peptide synthesis the amino acid corresponding to the carboxyl end of a peptide chain is attached to a heterogeneous matrix and the chain is then extended toward the amino end by stepwise coupling of activated amino acid derivatives. Filtration and washing remove the soluble excess reagents and byproducts. After the synthesis of the desired peptide chain, the peptide is removed from the heterogeneous support and purified. The plausible advantages underlying the solid phase peptide synthesis are:

— the reactions can be driven to completion through the use of excess soluble low molecular weight reagents;
— mechanical losses can be avoided by retaining the peptide-polymer beads in a single reaction vessel throughout the synthesis;
— the physical manipulations of the heterogeneous system can be automatized.

In addition to its insolubility and rigidity, a suitable carrier for solid phase stepwise synthesis should be capable of functionalization to a relatively high degree. The insolubility of the polymer beads facilitates filterability, and the rigidity provides good mechanical characteristics. However, the matrix should be flexible to a certain extent in order to be swelled in suitable solvents. The functional groups in the polymer should be easily accessible either in the rigid or in the swelled form, to the reagents and solvents. Improvements in the accessibility to reagents and solvents are sometimes achieved by grafting of the reactive functional groups to the polymer backbone by a long handle or a spacer arm. The functionalized macromolecular support should undergo straight-forward reaction with the reagents and be free of any side reactions. The macromolecular support should also be physicochemically compatible with the bound substrate, reagents and the solvents used for effective reactions to occur. The regeneratability by a simple, low-cost, high-yield reaction is another desirable qualification of a good polymeric support for peptide synthesis.

A wide variety of polymeric supports have been investigated giving due considerations to the above criteria. Polystyrene has been the most widely used polymer for this purpose, because of its commercial availability, ease of functionalization and good mechanical characteristics. However, limitations like the nonequivalence of functional groups and physicochemical incompatibility with the substrate observed in the case of this support have led to many strategical improvements in the solid phase methodology. In the following sections we discuss very briefly the basic features and the intrinsic problems of the polystyrene-based peptide synthesis.

### 2.1.1 Polystyrene Supports in the Solid Phase Method

The ease of incorporation of various functional groups, good mechanical characteristics and commercial availability are the factors which make the cross-linked polystyrene a commonly used macromolecular support for synthetic purposes. In the original Merrifield method, the C-terminal N-protected amino acid is bound by its carboxyl group to chloromethylated, cross-linked polystyrene through a benzyl ester linkage. Such an ester bond is stable during the various reactions employed in the different stages of the peptide synthesis. In addition to chloromethylated polystyrenes, a number of other supports incorporating functional groups like phenacyl, hydrazyl, acylsulfonyl, benzhydryl, aminomethyl etc. have also been used in the solid phase peptide sythesis [11]. A number of these resins, which are 1–2% cross-linked, are now commercially available. These products meet uniform specifications with respect to properties like particle — size and shape, degree of cross-linking and functional group capacity.

The chemical steps involved in the Merrifield's peptide synthesis using chloromethylated polystyrene is outlined in scheme 1. After the incorporation of the first amino acid to the polymer through a benzyl ester linkage, the terminal amino group is deprotected under conditions which do not cleave the resin-amino acid ester bond.

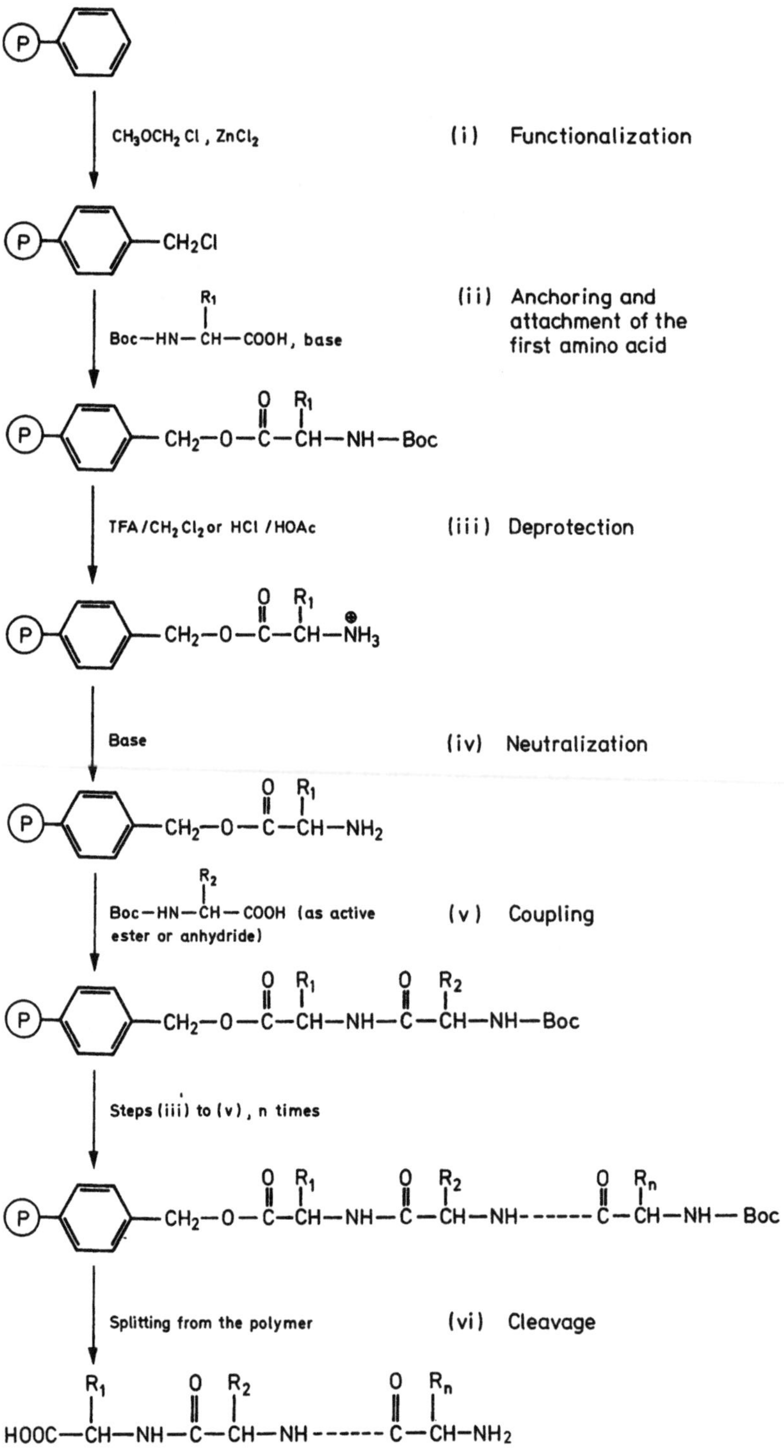


*Scheme 1.* A Typical Outline of the Solid Phase Peptide Synthesis.

Then another N-protected amino acid is coupled to the free amino group of the polymer-bound substrate using the dicyclohexylcarbodiimide activation or the active ester method. The N-deblocking and coupling steps are repeated until the desired sequence is formed. Finally the resin-peptide bond is split by a suitable acid cleavage reaction with HBr—AcOH, trifluoroacetic acid or HF. This results in a simultaneous N-deblocking and deprotection of most of the side-chain functionalities.

The problems involved in all of the above steps have been the subject of extensive investigations, and many significant and novel adaptations of the above basic strategy have been effected depending upon the case in hand. The use of various anchoring groups of different stability between the polymer support and the first amino acid particularly facilitated the attachment of the first residue and the final cleavage of the finished peptide [11].

### 2.1.2 Analytical Control in the Stepwise Synthesis

The purity of the final product-peptide obtained from a Merrifield stepwise synthesis depends upon the extent of completion of the deprotection and coupling reactions. Even very small deviations as low as less than 1% from completion could result in a statistical distribution of failure sequences and truncated peptides along with the target peptide [16]. Therefore the efficiency of the deprotection and coupling steps should be quantitative, or at least better than 99% for the synthesis of reasonably pure longer peptide chains. Thus, in order to achieve an effective synthesis of the desired peptide, it is highly necessary to have rapid analytical control of the coupling and deprotection stages in each synthesis cycle. Innumerable analytical methods have been put forward for this purpose and this topic has been the subject of a number of reviews [17].

The existing analytical methods for the solid phase peptide synthesis determine the substitution level of the growing peptide chain bound to the support or determine the endpoints of the deprotection or coupling reactions. The methods which analyze the consumption of a soluble reagent or release of a soluble byproduct are less sensitive than the methods which determine the unreacted polymer-bound peptide chains because of the relative error factors involved. Thus, continuous-flow methods which reversibly measure all the unreacted amino groups in the total batch soluble species do not provide a precision greater than 1%. The amount of free amino groups remaining after coupling is determined either by the non-destructive methods which reversible measure all the unreacted amino groups in the total batch of the resin or by the destructive methods which irreversibly measure the free amino groups in a definite amount of the resin.

Ultraviolet spectrophotometric analysis of the incorporation or deblocking of the α,α-dimethyl-3,5-dimethoxybenzyloxycarbonyl (Ddz) group in solid phase synthesis has been used by Birr for the analysis of the coupling and deprotection reactions respectively [18–20]. Similar spectroscopic properties of the N-protecting groups have been used for the analysis in the solid phase synthesis involving Nps-amino acids [21,22] and Bpoc-amino acids [23]. In all these methods the uptake or release of a soluble reagent or byproduct is analyzed.

Bayer et al. used mass spectrometry to detect the formation of failure sequences in the solid phase method [16]. Mass spectrometric dilution assay coupled with Edman degradation was employed for the determination of uncoupled amino component by Weygand and Obermeier [24].

$^{14}C$-labelled and $^{35}S$-labelled phenylisothiocyanates have been used in place of the usual phenylisothiocyanate in Edmann degradation and the free amino groups have been determined by measuring the labelled phenylthiohydantoin released from the resin [25,26]. The reaction of fluorescamine with primary amino groups to give fluorescent products that emit at 475 nm when excited at 390 nm [27] is made use of in the fluorometric assay of unreacted amino groups developed by Felix and Jimenez [28]. This method can be directly applied to the resin-bound peptide and upto 0.1% free amino groups can be detected in this case.

The determination of the unreacted amino groups in peptide-containing resins by direct titration with perchloric acid in glacial acetic acid or in acetic acid/methylene chloride is a rapid and very sensitive method introduced by Brunfeldt et al. [29,30]. This potentiometric titration using glass and calomel electrodes inserted directly into the reaction vessel has been used to measure the extent of both deprotection and coupling and the method can be automated. The development of these sensitive heterogeneous chemical monitoring techniques helped the automation of the stepwise solid phase peptide synthesis very much.

### 2.1.3 Mechanization and Automation

The most attractive feature of the solid phase technique for peptide synthesis is its amenability to mechanization and automation. Due to its suitable mechanical characteristics, the cross-linked polystyrene solid support significantly simplifies the manipulations involved in the reagent transfer, filtering and washing. Merrified and Stewart [31] designed the first mechanized assembly in 1965 in which the above procedures were manually operated. This original machine consisted of a glass reaction vessel with a sintered glass filter, a liquid-handling system for the manipulation of solvents, and a programmer to control the sequence of operations [32]. Since the introduction of this apparatus a variety of similar devices have been constructed by various investigators by making use of the same basic principle, but mainly differing in the design of the programmer. Brunfeldt has reviewed the characteristics of these improved mechanized assemblies for solid phase peptide synthesis [33].

The progress in the development of improved monitoring techniques for the heterogeneous synthetic reactions involved in the solid phase method provided the feedback control which was necessary for the full automation of the peptide synthesizers. A monitoring system, which is based on the titration of the unreacted polymer-peptide chains with picric acid [34,35], on coupling with a peptide synthesizer provided an automatic feedback [36]. This feedback is to implement the next step in the synthesis if the level of the unreacted peptide chains is below the acceptable preset value, or otherwise to repeat the last step.

The monitoring system based on the potentiometric titration of resin-bound unreacted amino groups [29,30] has helped the development of much advanced fully automated peptide synthesizers [37,38]. This computarized system evaluates automatically the synthetic data and decides whether to proceed to the next step, to repeat the last step, or to stop the synthesis. Very recently Edelstein et al. [39] designed and implemented a microcomputer system to control all the manipulations in solid phase peptide synthesis. In this assembly, reportedly, when standard synthetic programmes are used, the operator need only type in the amino acid sequence and give the "start" command to initiate the synthesis.

A number of automatic peptide synthesizers are commercially available now [40]. However, in spite of all these mechanical and electronical advances in the automation of solid phase synthesis, as Barany and Merrifield state in their review on the solid phase peptide synthesis [11], the limiting factor continues to be the chemistry of the process and automation can reach its full potential only when the assorted chemical difficulties are under control.

## 2.2 Incompatibility of Polystyrene Supports

In the stepwise synthesis of peptides using cross-linked polystyrene supports, the rate of incorporation of a particular amino acid residue has been found to decrease with increasing chain length in a number of instances [41–43]. These occurrences have been attributed to steric hindrance at the various functional sites on the heterogeneous network. Sheppard investigated the origin of this steric hindrance and its sudden and unpredictable onset on the reactivity and physicochemical characteristics in the case of the cross-linked polystyrene-bound peptides [44]. These studies suggested that the physicochemical incompatibility of the polystyrene matrix with the attached peptides is the factor responsible for the undesired influences of the solid support on the synthetic manipulations [44,45].

The polystyrene support with the growing peptide chains on it can be considered as a graft copolymer with a hydrophobic backbone and hydrophilic pendant peptide chains. In this graft copolymer, the backbone and the pendant chains have entirely different polarities. As the chain elongation proceeds, the polarity of the peptide chain changes. In a non-polar solvent, one can envisage collapse of the pendant peptide chains while the polystyrene backbone remains largely extended; in a polar, solvating medium, the hydrocarbon backbone may collapse, entangling with it the shorter peptide chains [44].

Thus, in the non-polar medium, due to the collapse of the otherwise partially extended peptide chain within itself, one can rationalize the steric hindrance; in the polar medium, due to the collapse of the otherwise extended polystyrene backbone, some of the peptide chains can be visualized as "buried" and hence hindered to the approach of reagents and solvents. Thus, based on these models of polystyrene-bound peptides in different environments (Fig. 1), an ideal situation would be that where both the polymer and the peptide chains are extended. This is likely to be attained most easily if the polymer and peptide are of comparable polarities and are placed in a good solvating medium. This is in contrast to the polystyrene case where the macromolecular support is a pure hydrocarbon physicochemically quite dissimilar to the peptide chain being synthesized [44].

The physicochemical incompatibility of the polystyrene support with the attached peptides is an impediment to the mass transport of reagents, effective solvation of the polymer as well as peptide and to the enhanced rates of coupling and deprotection reactions [45]. Many attempts have been made to overcome this difficulty in the solid phase peptide synthesis. Some of these attempts resort to the development and

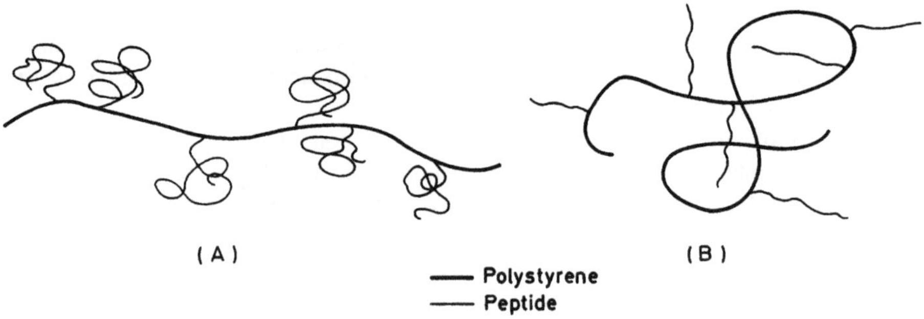


**Fig. 1.** Hypothetical representation of polystyrene-bound peptides in (A) non-polar solvent and (B) polar solvent

application of appropriate protecting groups in order to make the peptide more physicochemically compatible with the polymer support[46,47]. A more general approach to solve the problem would be the development of supports which are structurally similar to the backbone structure of the peptide; this could overcome the dynamic changes in the solvation of the polymer and the bound substrate. The proposed use of polypeptide supports for peptides and oligonucleotides[48,49] appears to be more of fundamental interest. The design of synthetic, polar polymeric supports like the polyacrylamide supports and polyacrylylpyrrolidine supports is a significant advance and this holds promise for substantial improvements in the solid phase peptide synthesis.

## 2.3 Polyacrylamide-Type Supports for Peptide Synthesis

Based on the principle of the equal and simultaneous solvation of the polymer and the bound peptide chains in different solvents, Sheppard and coworkers developed a number of polyacrylamide-type supports for solid phase peptide synthesis[50–55]. In this case, the crosslinked polymeric support, in addition to possessing the good mechanical characteristics like polystyrene, is much more structurally related to the peptide than in the case of polystyrene. The polar polyacrylamide support in this case is prepared by the emulsion copolymerization of a mixture of dimethylacrylamide (*1*), ethylenebisacrylamide (*2*) and acryloylsarcosine methylester (*3*), initiated by ammonium persulphate[51].

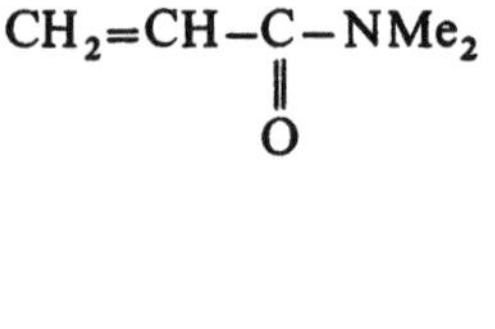

*1*

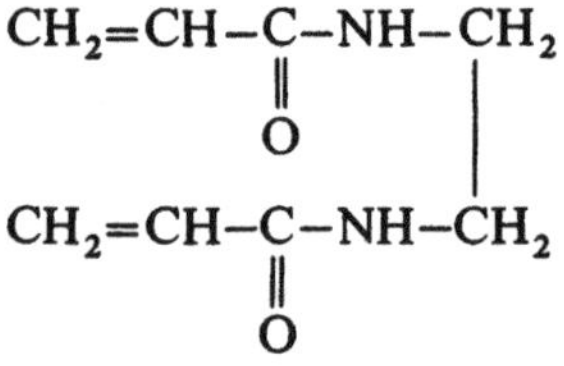

*2*

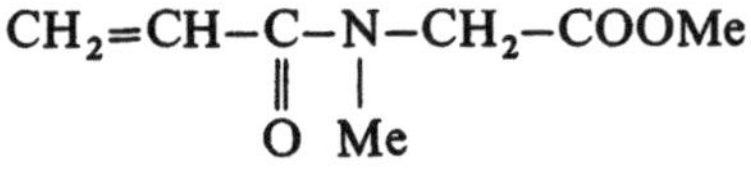

*3*

The acrylamide *1* serves as a basic monomer, *2* as a cross-linking agent and *3* as the functionalizing substituent. The resulting polyethylene-polyacrylamide support with the structure *4* can be obtained as beads with good solvation properties. Thus, 1 gram of this resin swells to a volume of 20 ml in a wide range of solvents like water, methanol, dimethylformamide, pyridine and methylenechloride.

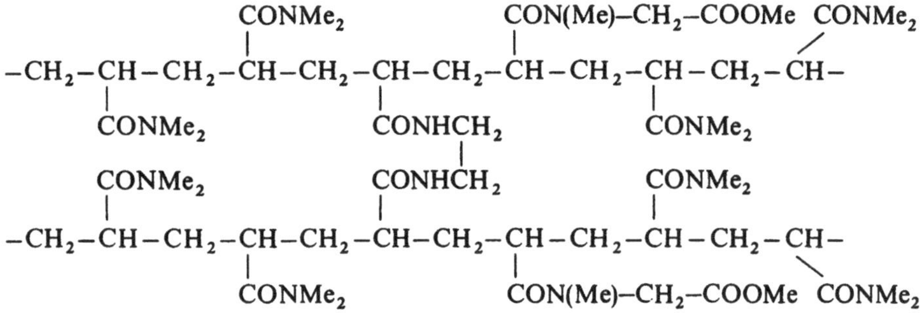

*4*

Complete conversion of methoxycarbonyl groups into primary amino groups was achieved by treatment with excess of ethylenediamine overnight at room temperature. The degree of functionalization of this resin sample as judged by both sarcosine content and subsequent amino acid incorporation was 0.35 mequiv./g. This resin has been used successfully for both peptide and oligonucleotide synthesis applying the solid phase methodology.

Earlier Sheppard and coworkers used *N*-acryloyl-*N'*-(butyloxycarbonyl-β-alanyl)-hexamethylenediamine (*5*) as the functionalizing substituent in place of the acryloylsarcosine methyl ester (*3*) [51-54].

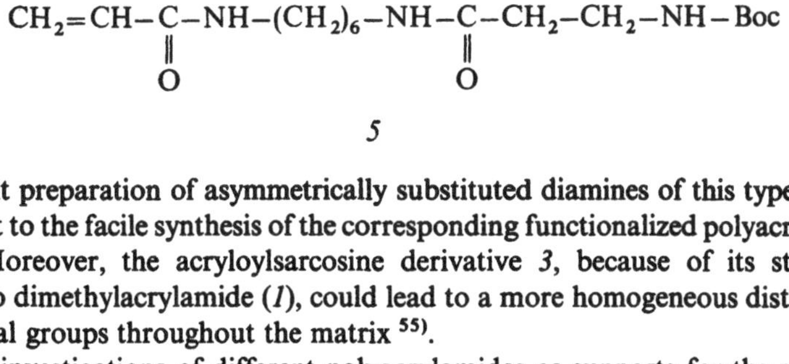

*5*

The difficult preparation of asymmetrically substituted diamines of this type was an impediment to the facile synthesis of the corresponding functionalized polyacrylamide support. Moreover, the acryloylsarcosine derivative *3*, because of its structural similarity to dimethylacrylamide (*1*), could lead to a more homogeneous distribution of functional groups throughout the matrix [55].

Detailed investigations of different polyacrylamides as supports for the synthesis of peptides and oligonucleotides helped Atherton et al. to define the following requirements for polyamide supports to be suitable for the solid phase peptide synthesis [53].

- presence of a high proportion of tertiary amide groups which give good swelling characteristics in suitable polar organic solvents. Such tertiary amide groups prevent intrachain hydrogen bonding which otherwise results in polymers permeated only by water. Thus, crosslinked polyacrylamide which swells only in water and in very highly polar media finds little application in peptide synthesis even though it had been proposed for this purpose earlier by Inman and Dintzis [56].

— stable crosslinks, derived from bisacryloyl derivatives of ethylenediamine or higher amines, but not from the more usual methylenebisacrylamide (the latter is not stable to repeated treatment with acids).
— a low degree of functionalization (0.2 to 0.5 mequiv./g) with suitably protected amino groups which may be associated with an internal reference amino acid (to facilitate accurate analytical procedures).
— the need for minimum chemical manipulation of the polymer subsequent to its preparation.

A number of model peptides and protein sequences have been prepared by the solid phase method using these polar polyacrylamide supports. These include Acryl Carrier Protein [53], human-β-endorphin [54], Substance P [57], Pancreatic Trypsin Inhibitor [58], and minigastrin analogs [58]. In many of these cases the yields were higher compared to the synthesis using the polystyrene resins [59,60]. These polar supports have been used for oligonucleotide synthesis [61] and solid phase protein sequencing [62]. Very recently this resin has been incorporated within the pores of rigid macroporous inorganic particles like kieselguhr to produce a physically supported gel polymer which find application in continuous flow solid phase peptide synthesis [63]. This polydimethylacrylamide-kieselguhr composite support was prepared by the emulsion copolymerization of the corresponding monomer, crosslinking agent and functionalizing agent [51] in the presence of fabricated kieselguhr particles.

## 2.4 Poly(*N*-acrylylpyrrolidine) Support

Giving due considerations to the chemical nature and to the topographical structure of the target polymer matrix, Walter and coworkers designed a macromolecular support, Poly(*N*-acrylylpyrrolidine) (PAP), which has the mechanical characteristics and compatibility requirements necessary for the application as a solid support for peptide synthesis [64,65]. The PAP resin in bead form was prepared by the reverse phase suspension copolymerization of *N*-acrylylpyrrollidine (*6*), ethylenebisacrylamide (*2*) (crosslinking agent) and N-acrylyl-1,6-diaminohexanehydrochloride (*7*) (functionalizing agent).

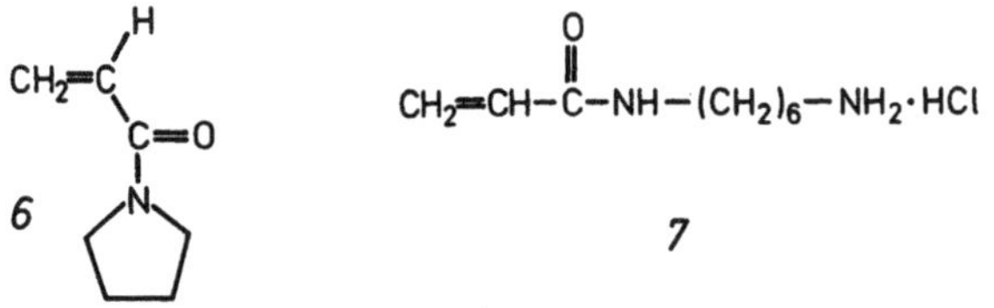


*6* *7*

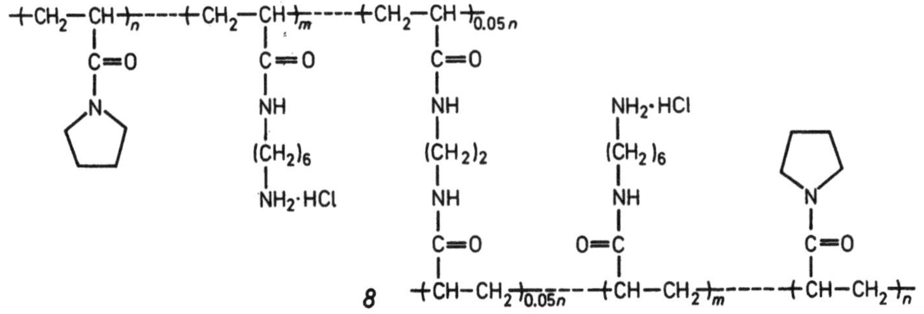


*8*

The PAP resin (*8*), like the polyacrylamide resins, overcomes the general problem of peptide synthesis on cross-linked polystyrene, namely the different solvation behaviour of the growing polar peptide chain and the lipohilic styrene matrix. This resin with a degree of crosslinking 4.4 and 0.7 mequiv. of amino group per gram shows remarkable promise as a support for solid phase peptide synthesis. In its protonated or acylated form, the resin exhibits favourable swelling characteristics in a variety of solvents ranging in polarity from water to methylenechloride. Comparison of the swelling properties of Boc-Gly-Polystyrene-1 %-DVB resin and Boc-Gly-PAP resin (specific volumes of the dry resins, 1.5–1.6 ml/g) showed useful swelling of the Merrifield resin in three solvents (DMF, 5.2; $CH_2Cl_2$, 7.6; chloroform, 9.5), whereas the PAP resin swells well in many solvents including methanol (6.6), ethanol (7.3), isopropanol (6.5), trifluoroethanol (11.2), dimethylformamide (5.2), $CH_2Cl_2$ (7.5), $CHCl_3$ (8.7), acetic acid (8.5) and even in water (6.2). These especially favourable swelling properties of PAP allows the use of most of the techniques developed for the solid phase method as well as the development and employment of new techniques that rely on aqueous or other polar solvents (eg. liquid $NH_3$), the use of which are precluded by the poor swelling properties in such solvents.

The feasibility of the attachment and detachment of amino acids and peptides to and from the PAP resin has been demonstrated by using S-carbamoyl, dinitrophenyl and benzylester polymer-to-peptide bridging groups. By using these bridging groups the bidirectional solid-phase syntheses [66–68] of deaminooxytocin [65], Thyrotropin-Releasing Hormone (TRH) [65], Glu- and Gln-containing model peptides [65], and ($Asp^5$)-arginine vasopressin [69] have been carried out.

## 2.5 Segment Condensation on Solid Polymeric Supports

The condensation of purified, soluble protected peptide segments to a resin-bound peptide reduces the difficulties resulting from the accumulation of failure sequences on the resin in the stepwise solid phase peptide synthesis. The method is essentially the same as the stepwise approach except that the soluble components are protected peptides instead of protected amino acids. As in the stepwise method, the amino component peptide is resin-bound, while the carboxylic component peptide is kept in solution. Shortly after the introduction of the Merrifield's stepwise method, Weygand and Ragnarson reported the use of phenacylpolystyrene resin for coupling peptide segments on solid support [70].

Although small and simpler peptide segments can be coupled by the active ester method in the solid phase method, azide coupling as in the solution method is more suitable for coupling larger peptide segments. The soluble protected C-terminal peptide component can be prepared in pure form by the stepwise solid phase method. This strategy was used to assemble a number of medium-sized peptides from soluble C-terminal segments and resin-bound N-terminal segments [71]. The protected peptide hydrazide can be directly obtained from the stepwise solid phase synthesis by making use of suitable resins, and this segment can then be condensed with the resin-bound N-terminal segment, *via* the azide coupling [72–74]. Birr has investigated in detail the potential of the segment condensation on the polymeric supports and synthesized a number of biologically active proteins by this method [75,76].

The advantages of the solid phase condensation approach include
1. easy removal of the excess, soluble peptide component,
2. lowering of the byproduct — failure sequences — distribution because of the smaller number of transformations, compared to the stepwise method and
3. the possibility of more effective separation and purification of the final sequence due to the pronounced differences in molecular weights among the failure sequences on the support themselves and the target sequence.

However, the method has some *drawbacks*, compared to the solution method, arising from the polymer-bound nature of one of the segments. The complete solvation of both of the components, which is necessary for the effective condensation, cannot be easily achieved and this results in a reduced reactivity of the polymer-bound N-terminal component. Consequently, the usual racemization on the C-terminus of the soluble peptide segment during the coupling is further enhanced in the polymer phase reaction. Methods like azide coupling and dicyclohexylcarbodiimide/*N*-hydroxysuccinimide activation (which cause about 5% racemization even in the solution case) result in increased racemization because of the longer reaction periods which are necessary for effective condensation with the polymer-bound segment. But 1-hydroxybenzotriazole in combination with dicyclohexylcarbodiimide activation does not favour racemization even in extended reaction periods in the polymeric phase just as in the solution case. This has led to a number of successful attempts in the segment condensation on polymeric supports [75,77].

Biologically important peptides and proteins which have been synthesized by the solid phase segment condensation include the Bovine basic inhibitor [78,79], ACTH sequences [80], insulin and proinsulin [81], and glucagen [82]. A number of sequential polypeptides have also been synthesized by this method [83,84]. Segment coupling on solid supports with attachments at the side chain functionalities has been adapted for the synthesis of glutamine peptides. This procedure was employed for the synthesis of the C-terminal peptide of human proinsulin and serum thymic factor [85]. Sheppard's polyacrylamide approach has also been extended to the solid phase segment condensation approach as illustrated by the synthesis of the gastrin sequences [86].

## 2.6 Multipurpose and Multidetachable Polystyrene Supports

The recent introduction of multidetachable polystyrene resin supports which make use of selectively cleavable anchoring groups at different positions provides much more flexibility and adaptability for the synthetic design in the solid phase peptide synthesis [87–90]. This versatile group of resin supports contains two orthogonal[1] ester bonds, which allow the selective cleavage of the individual bonds by a variety of reagents. These supports contain an attachment, peptide—O—X—O—Y resin, with two orthogonal ester bonds, bond *a* (peptide—O—X) and bond *b* (X—O—Y), separated by a suitable spacer X (scheme 2). By using suitable reagents one or the

[1] An orthogonal system in the functional group protection strategy is defined as a set of completely independent classes of protecting groups. In a system of this kind, each class of groups can be removed in any order and in the presence of all the other classes. In other words, this denotes a combination of chemoselectively removable protecting groups [91,92].

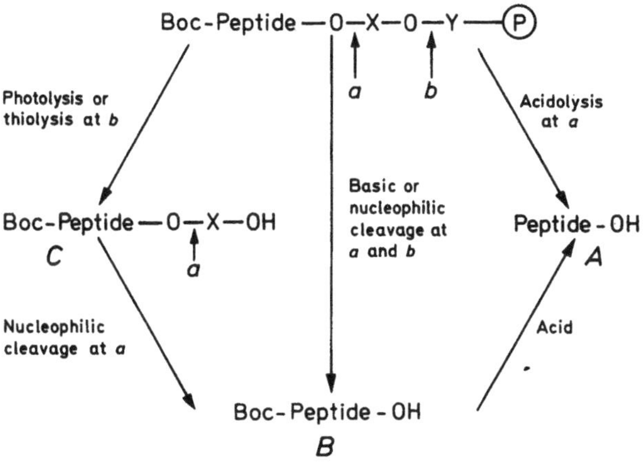


*Scheme 2.* Multidetachable Resin Supports for the Solid Phase Peptide Synthesis.

other ester bonds can be cleaved selectively under conditions where the second bond is stable. After the synthesis, the final peptide can be obtained in a free form (A), in a protected form (*B*), which is suitable for segment condensation, or with the removable spacer X still attached to the carboxyl terminus (form *C*); this intermediate form *C* can be reattached to another suitable resin support for further stepwise or segmental elongation. In this system, a mild interconversion from one form to the other is also possible [88,89].

The supports *9* and *10*, which were synthesized starting from cross-linked chloromethylpolystyrene, contain an acid-labile benzyl ester bond (bond *a*) and a photolabile *o*-nitrobenzyl or α-methylphenacyl ester bond (bond *b*). The former is stable to photolysis and the latter are stable to acidolysis; the 4-oxymethylphenylacetyl group serves as the spacer X (Scheme 2) and this permits basic or nucleophilic cleavage.

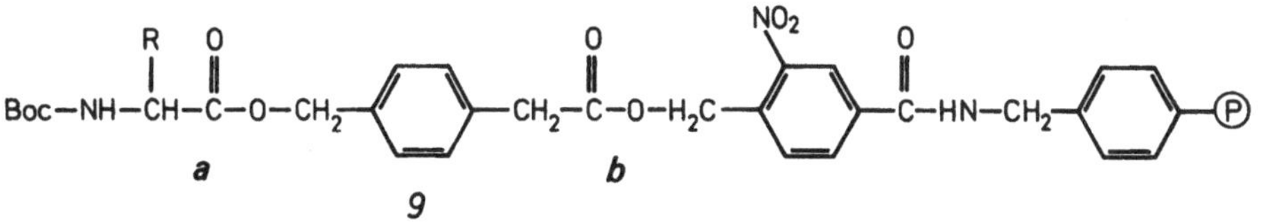


*9*

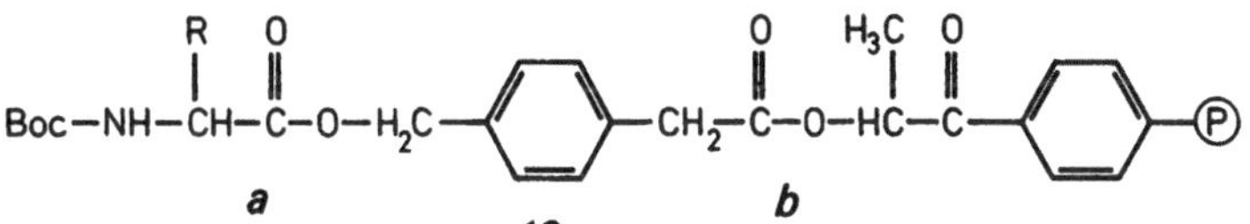


*10*

The use of the multidetachable resin supports has been observed to offer considerable degree of freedom and versatality for the synthetic schemes. The cleavage of the peptide derivatives from these resins is possible under very mild conditions and does not lead to any side reactions which are harmful to the sequences.

Another important advantage of these resins is that they provide a route for the preparation of the peptide derivatives with the 4-oxymethylphenylacetyl handle (eg. peptide derivative of the type *11*) which permit reattachment to a suitable resin for further synthetic manipulation. The application of these resins has been illustrated by the synthesis of Leu-enkephalin and angeotensin II [89].

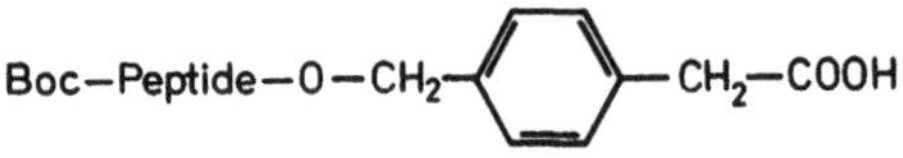


*11*

## 3 The Polymeric Reagent Method

The polymeric reagent method of peptide synthesis introduced by Fridkin et al. [93,94] is based on the use of insoluble polymer-supported amino acid active esters as the carboxyl component for the coupling to the soluble amino terminal component. In this method, the peptide formed remains in solution; it can be isolated, purified and its purity checked before use in the subsequent reactions in the stepwise method. This is in contrast to the Merrifield method where the peptide is cleaved from the resin only after the entire sequence of the synthetic steps are completed. Thus the possibility of the isolation and purification of the intermediate peptides eliminates the heterogeneity of the final product in the polymer reagent method; the coupling efficiency in this instance can also be increased by using excess of the insoluble supported reagent, and the excess reagent after coupling is removed by filtration.

The polymer reagent method has been used in peptide synthesis by other investigators also [95–97]. The method has been applied to the synthesis of a number of medium-sized peptides and biologically active protein sequences [98].

In this section, we briefly discuss the basic features, specific applications scope and limitations of the method in order to illustrate its significance in the development of other synthetic strategies for peptides using polymeric supports.

### 3.1 Polymeric Active Esters in Peptide Synthesis

In a polymeric active ester, the acyl component A is bound covalently *via* an active ester link to the polymeric carrier Ⓟ. Such reagents, Ⓟ—A, readily react with the nucleophile B — in peptide synthesis, the amino component — to give the product A—B. In order to obtain high yields of the peptide A—B, an excess of Ⓟ—A, which can be later removed from the reaction mixture by filtration or centrifugation, is used.

For further elongation of the peptide chain, successive coupling reactions with other polymeric active esters may be carried out until the desired sequence is obtained. The polymeric active esters are prepared by the attachment of the N-protected amino acids or peptides to a suitable insoluble polymer. Potentially useful polymeric supports in this case should carry a free hydroxyl function to which the carboxyl group of the amino acid derivatives may be coupled. Crosslinked poly(4-hydroxy-3-

nitrostyrene) (*12*), *N*-hydroxysuccinimide containing polymers (eg. *13*) and cross-linked polystyrene-bound 1-hydroxybenzotrialzole (*14*) are the typical, most commonly used hydroxyl group containing macromolecular supports for this purpose.

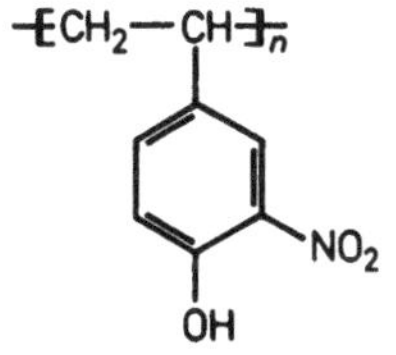

*12*

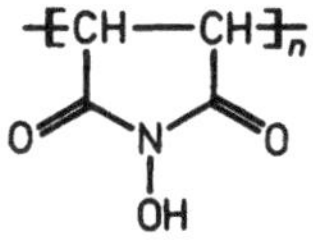

*13*

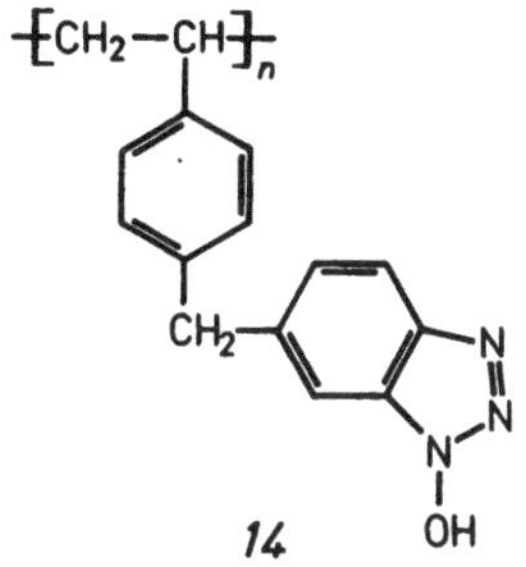

*14*

The utilization of the polymeric active esters in the step-by-step peptide synthesis is illustrated in scheme 3 with a typical example of the polymer-bound *o*-nitrophenol.

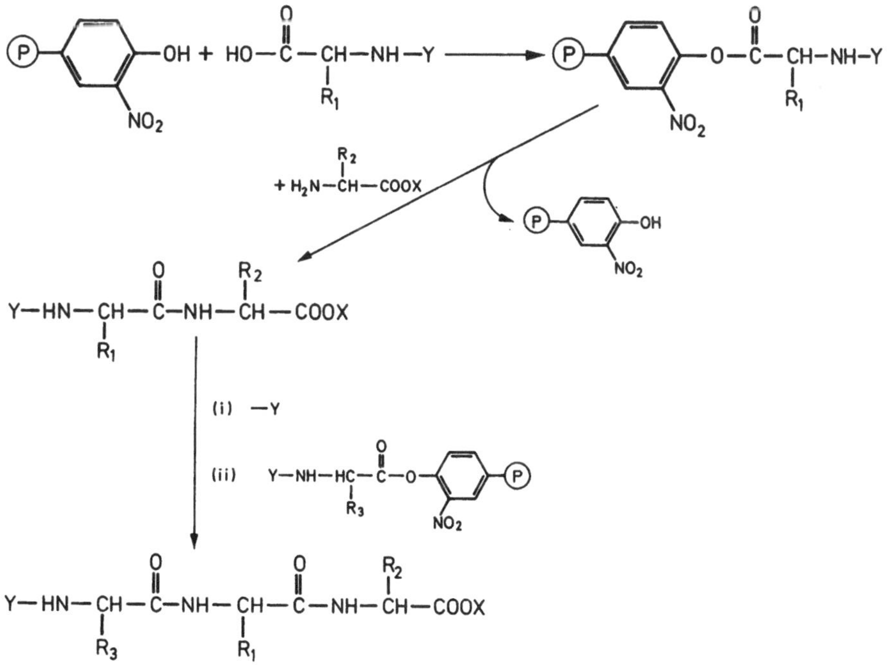


*Scheme 3.* A Typical Synthetic Cycle in the Polymer Reagent Method.

The polymeric reagent (containing generally 1.25 to 3.0 equivalents of the desired amino acid active ester) suspended in an organic solvent like methylenechloride, dimethylformamide or chloroform is allowed to react with the soluble carboxyl-protected amino acid. The resulting amino- and carboxyl-blocked dipeptide is then isolated by filtering off the spent polymeric reagent and evaporating the reaction solvent. Synthesis can be carried out further through the selective removal of the amino protecting group and acylation of the newly exposed free α-amino group by the appropriate polymer attached activated N-blocked amino acid. The cycle is repeated till the desired peptide is obtained.

Fridkin ascribes several features as prerequisites for ideal polymeric active esters for use in peptide synthesis: [98]

1. Ability to retain the mechanical stability (integrity of polymeric particles over many repetitive synthetic and reloading operations).
2. Good swellability in the organic solvents which are commonly used in peptide synthesis.
3. Accessibility of the active ester function to the incoming amino component to achieve quantitative acylation.
4. Minimal dependence of the acylation reactions on steric hindrance.
5. High reactivity.
6. Possibility of loading relatively large amounts of active esters.
7. Stability during storage.

## 3.2 Racemization in the Polymer Reagent Method

Detailed investigations of Fridkin et al. suggest that racemization is rather insignificant in the stepwise synthesis of peptides in which a single N-protected amino acid is transferred from a polymeric donor to a free α-amino group acceptor in a carboxyl-protected amino acid. Thus, for example, it was observed that properties like optical rotations, melting points and chromatographic mobilities of various peptides synthesized by the polymeric reagent method were identical with the values of those prepared by conventional and "racemization free" procedures [99–102]. The synthesis of the dipeptide Z-L-Val-L-Val-OBzl by the reaction of a polymeric *N*-hydroxysuccinimide-Z-L-Val ester with H-L-Val-OBzl at 70 °C has been observed to proceed without any racemization [99]. A comparison of the racemization tendencies of the polymeric *N*-hydroxysuccinimide and *o*-nitrophenyl esters of Z-Gly-L-Ala showed that in the former case, coupling with H-L-Leu-OBzl led to 6.4% racemization of the alanine residue whereas in the latter case the racemization was as high as 37.4% [103].

## 3.3 Mechanization of the Polymer Reagent Technique

Mechanization and automation are possible in the synthesis of peptides using solid polymeric active esters also. By passing a solution of the amine component in a suitable solvent through a column packed with the solid polymeric activated carboxyl component, mechanization could be effected. The product, the protected peptide, which is in the eluent, is then N-deprotected, and the product in solution is passed through yet another column containing a suitable reagent, to effect further elongation. Thus, for example, Wieland and Birr have used columns packed with active esters derived from N-blocked amino acids and cross-linked *p,p'*-dihydroxydiphenylsulfone (*15*) to synthesize several peptides in nearly quantitative yields [104].

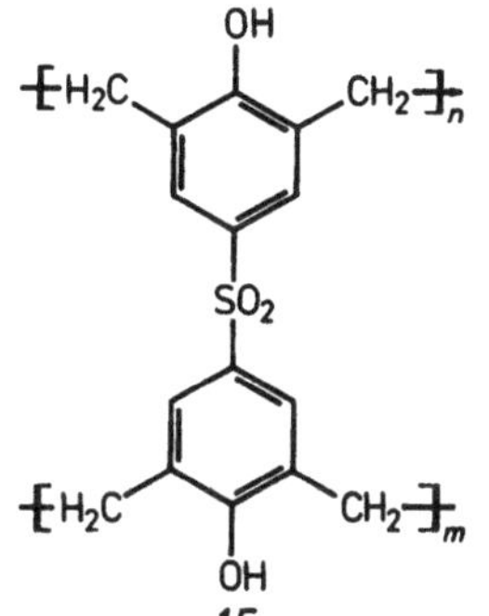

*15*

Stern et al. developed a batch-processing approach to mechanized peptide synthesis using polymeric active esters [105]. The procedure consisted of the following three consecutive steps without isolation of the intermediate compounds:

1. coupling of a carboxyl-protected amino acid with a polymeric active ester to yield the N- and C-blocked dipeptide,
2. removal of the N-protecting group, and
3. neutralization of the protonated form with a polymer-bound amine and subsequent treatment with a polymer-bound active ester to afford the amino- and carboxyl-blocked tripeptide.

The entire procedure was conducted in two round-bottom flasks in alternate succession of couplings and deprotections. This simple and smooth batch-processing technique has been used to prepare peptides corresponding to human ACTH (19–24) and human ACTH (15–18) in 63% and 70% overall yields, respectively [105].

## 3.4 Specific Applications, Scope and Limitations

The polymeric active ester method has been used successfully for the preparation of several small- to medium-sized peptides in very pure form; the potentiality of the method has also been illustrated by the synthesis of a number of biologically active protein sequences like bradykinin [106], thyrotropin-releasing hormone [107], ACTH sequences [105], and LH—RH [108] in good overall yields.

Another important application of the polymer reagent method is in the synthesis of cyclic peptides. In this approach, the amino-protected linear peptide (*16*) is converted into the corresponding polymeric active ester (*17*) by coupling it to a cross-linked hydroxyl function containing polymer (eg. *12*); on deprotection of the amino group and subsequent neutralization, the cyclic peptide (*18*) is obtained in good yields [109, 110]:

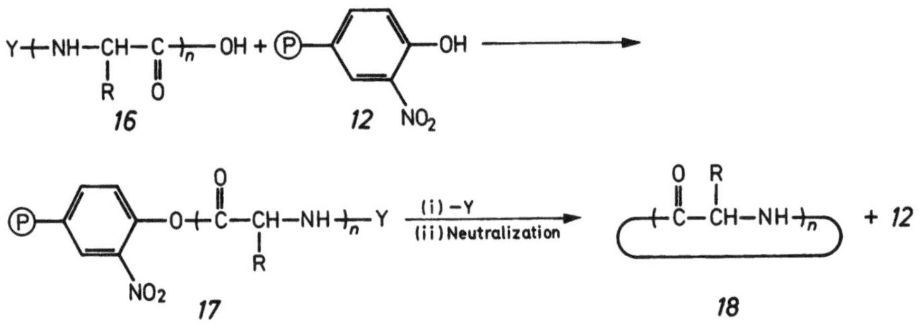

The facilitation of this intrapeptide reaction has been explained on the basis of the possibility that in the rigid polymer matrix each peptide molecule is isolated from others and this would bring in the condition of infinite dilution necessary for the intramolecular cyclization (for a discussion of site-isolation in solid phase reactions see section 4.3). The method has been applied to the synthesis of several cyclic peptides in good yields.

In contrast to the Merrifield method, in the polymer reagent method, the growing peptide remains in solution; therefore during the later stages of the synthesis, the solubility of the oligopeptide is a decisive factor in this method. With increasing chain length of the peptide the solubility may decrease and this eventually will obstruct further synthetic manipulations. Another difficulty when one considers the synthesis of larger peptides by this method is the possibility of steric hindrance which can be expected from the acylation of a polypeptide with the polymeric reagent.

However, Fridkin's observations suggest that as long as the polypeptide to be acylated remains soluble in a suitable organic solvent, its coupling with the polymeric active ester occurs efficiently [98]. Thus, it was observed that both insulin and poly-benzyloxylcarbonyllysine, when dissolved in dimethylformamide, were quantitatively acylated by the polymeric N-hydroxysuccinimide active ester of Boc-Ala at rates comparable to those of low-molecular weight amines [111]. This points to the conclusion that decreased solubility is the only significant limiting factor in the utilization of the polymer reagent method for the synthesis of large polypeptides.

To approach this insolubility problem, Fridkin [98] suggests the combination of the polymer reagent method with the liquid phase synthesis using soluble polymeric supports or the use of appropriate solubilizing protecting groups. (See Section 5 for the application of soluble polymeric supports). A systematic analysis of the factors governing the reactivity, solubility, solvation and other physicomechanical characteristics of the peptide, polymer and the polymer-bound peptide is therefore of utmost importance for a proper utilization of all these synthetic strategies for peptides.

## 4 Reactivity in Polymeric Solid (Gel) Phase Reactions

The success of both the *stepwise synthesis of peptides* on a cross-linked polystyrene support (which involves the reaction of a low molecular amino acid derivative to a polymer-bound amino group) and the *polymer reagent technique* (where the growing peptide in solution is treated with a polymer-bound activated amino acid) depends on whether the reaction of the polymer-bound functional group with the reagents in solution will go to completion.

One of the main difficulties associated with the above types of polymer-supported syntheses is the non-equivalence of reactive groups attached to the polymer network. This may lead to inaccessibility of some sites to solvents and reagents. It might be expected that groups placed close to a cross-link might be less accessible to reagents in the continuous phase. Another difficulty with the cross-linked polystyrene resins for use in peptide synthesis is the drastic change in polarity of the network during the growth of the polypeptide chain. These difficulties originate from the steric effects in the cross-linked polymeric matrix, insufficient solvation of the growing peptide chain and conformational transitions occuring in the peptide.

## 4.1 Non-equivalence of Reaction Sites

For the polymer-based peptide synthesis it is neccessary that the reaction of the polymer-bound functional groups with an excess of the reagents in the solution phase should go to completion. In this connection, it is important to know whether the functional groups attached to a cross-linked polymeric network are kinetically equivalent.

Investigations by Morawetz [112,113] and coworkers showed that the reaction of anilino groups attached to partially swollen polymers with solutions of a large excess of acetic anhydride follows linear first order kinetics. But the introduction of cross-links into the polymer leads to deviation from first order kinetics. They observed that these deviations are insensitive to the degree of cross-linking, but quite sensitive to the nature of the polymer to which the anilino groups are attached [113]. The reason for this deviation is the changes in the local polarity of the cross-linked polymer system. In the absence of cross-links, the local polarity is averaged over the lifetime of the reaction by microbrownian motions, but this becomes impossible when the linear polymer is converted into a network structure.

Due to the nonreproducibility of the specific steric effects in the heterogeneous phase, reliable comparisons between the heterogeneous and homogeneous reactions are impossible. A number of investigations directed toward the optimization of the polymeric support for the solid phase peptide synthesis revealed a strong correlation between the physical nature of the solid matrix and coupling yields [114-119]. In many cases the chemical and physical non-equivalence of functional groups within the crosslinked matrix is a source of deviation from linearity in the kinetics of the reactions [117-119].

In the light of these observations, it is not surprising that comparative kinetic investigations in heterogeneous and homogeneous systems resulted in rather inconsistent data [120-122]. Thus, for example, Gut and Rudinger [121] found drastic deviations from linearity of the second-order rate constant for yields higher than 50%, whereas Erickson and Merrifield observed a close fit to second order kinetics upto 95% completion of reaction [11]. These results also indicate a drop in the reaction rates in the order of two- to five-fold for the DCC-mediated coupling in the heterogeneous system.

The fact that the reaction rates in solid phase synthesis are not drastically reduced, compared to the homogeneous reactions, indicates that the diffusion of the reagent into the polymeric matrix is not a limiting factor for the method. This has been confirmed by Andreatta and Rink [119] in kinetic studies on both cross-linked and linear polystyrenes. This means that the intrinsic problems of solid phase synthesis arise from deviations in the linear kinetic course in the final stages of reaction due to non-equivalence of functional groups.

## 4.2 Variables of Gel Preparation and Reactivity of Functional Groups

The reactivity of functional groups attached to a polymeric network is highly dependent upon the variables of gel preparation. Kau and Morawetz studied the role of the structure of the cross-linked network in deciding the reactivity of the functional group attached to the polymeric matrix [123]. They used as a model reaction, the solvolysis of nitrophenyl esters catalysed by crosslinked acrylamide polymers containing 10% by weight of 4-acrylamidopyridine residues. These investigations indicated that the degree of cross-linking and the topographical nature of the gel network affect the chemical reactivity of the attached functional groups.

### 4.2.1 Degree of Cross-linking

In the solvolysis reaction of nitrophenyl esters by differently cross-linked polyacrylamide resins containing 4-acrylamidopyridine catalytic sites, Morawetz and coworkers observed that an increase in the degree of cross-linking caused the rate of the reaction to increase, pass through a maximum, and then decrease [123,124]. Thus, two opposing factors must affect the reactivity of the gel. The reason for this behaviour could be that the catalytic pyridine groups are inaccessible to reagents in the continuous phase if the highly cross-linked gel is *only slightly* swollen and that the effective solvent medium is less favourable to the reaction if the swelling of the gel is *too large*. The latter point, namely the extent of solvation and swelling in gel-phase reactions, will be discussed in Section 4.4.

The extent of cross-linking is found to exert a striking influence on the reactivity factors in the stepwise coupling of amino acids on polystyrene supports. Letsinger and Kornet used low-swelling *popcorn polymers* of styrene-divinylbenzene containing as little as 0.01% cross-linking [125]. This low-density material is readily penetrated by solutes in solvents such as DMF, benzene, or pyridine even though it does not swell appreciably. Letsinger and Jerina [126] prepared representative specimens of 2% bead and 0.2% popcorn polymers bearing reactive groups such as nitrophenyl esters and investigated the kinetics of aminolysis. They found that the aminolysis occurs faster when bound to the popcorn polymer than when attached to the bead polymer.

The *macroporous or macroreticular resins*[2] are rigid highly cross-linked copolymers containing large pores and large effective surface. They provide another possible

[2] These polymers have an effective surface area several hundred times larger than that of suspension polymers. These resins usually react slower or to a lesser extent than their swellable counterparts. *Solvent swellable* resin beads usually contain 1 to 5% of cross-links. Their degree

way to increase the available sites and decrease the diffusion problems occuring in solvent-swellable (1–5% cross-links) polymers. These resins containing varying amounts of cross-links have been investigated by different groups for possible application in peptide synthesis by studying the rate of coupling at different stages. However, the results are contradicting in many cases so that a definite correlation between degree of cross-linking and rate of reaction cannot be drawn. Merrifield et al. observed that the extent of reaction decreased progressively as the peptide chain was lengthened beyond the range of 8–10 residues and they ascribe this to the limiting rates of diffusion between pores [11]. On the other hand, Losse observed no apparent difference between the efficiencies in peptide synthesis for a macroreticular resin and a solvent swellable resin gel [115]. He compared the synthesis of the eledoisin fragment, Boc-Lys-(Boc)-Phe-Phe-Gly-Leu-Met-$NH_2$, on an 8% cross-linked macroporous resin and on a 2% cross-linked gel resin bead. The coupling yield during each cycle was as high with the macroporous support as with the gel resin. A number of other peptide syntheses using different macroreticular resins have also been reported [129–131]. Frank and Hagenmaier measured the coupling yields during the synthesis of Gly-Val-Gly-Ala-Pro on several solid supports [116]. Peptide chains bound to a macroporous resin and several 2% cross-linked polystyrene resins coupled similarly, but all these coupled less efficiently than chains bound to a 1% cross-linked support.

Functional groups attached to solvent-swollen polymer chains exhibit free rotational motion as indicated by electron spin resonance rotational correlation times [132–134]. These studies using nitroxide spin labels covalently bound to polystyrene matrices indicated that the mobility of the substituent is a function of the cross-link density and degree of swelling. The rotational correlation time of nitroxide within 2% cross-linked beads was about 100 times shorter in dichloromethane or benzene than in ethanol, and 2–3 times longer than nitroxide bound to non-cross-linked polystyrene. The latter observation shows that the heterogeneous reaction involving 2% cross-linked polystyrene is 2–3 times slower than the same reaction in solution.

### 4.2.2 Topographical Nature of the Polymer Matrix

The topographical nature of a polymer matrix exerts a significant influence on the reactivity of the attached functional groups. Lloyd and Alfrey investigated the role of the topology of the gel network on chemical reactivity [135,136]. The topography of a polymer matrix is determined by the chemical nature of the monomers, the molar percentage of cross-link and the monomer dilution ratio (ratio of the amount of inert solvent to the amount of monomers by weight in the polymerization mixture). Morawetz and coworkers prepared a series of polyacrylamide gels containing pyridine

---

· of swelling is inversely proportional to the degree of cross-linking. Typically, 1 g of a 1% cross-linked polymer will swell to occupy a volume of 10 ml in benzene, while in the same solvent 1 g of a 5% cross-linked polymer will only swell to occupy 4 ml. Reactions in swollen 1 or 2% cross-linked polymer beads resemble those in dilute gels since the rates of reaction are of the same order of magnitude as those of reactions in solution [127]. *Popcorn polymers* are low-density polymers which contain only 0.1 to 0.5% of cross-links. They do not swell in any solvent and are easily penetrated by small molecules. They have a reactivity comparable to that of solvent swellable beads, but are often difficult to handle [128].

catalytic active sites by increasing the concentration of the cross-linking agent, while the monomer is diluted with an inert solvent during the suspension polymerization [123–124]. All these gels swell in a given solvent to the same extent; but the reactivities of the functional sites in the gels prepared at high dilution were much greater than those in others. These results show that the topological characteristics of the network structure play a definite role in controlling the energetic interaction between the cross-linked bead and the substrate in solution.

## 4.3 Intraresin Site Interactions

The concept of site-isolation in cross-linked resin-bound substrates has been a controversial subject [137]. The state of perfect site isolation in a polymer-bound substrate is an ideal situation in which there is no interaction between the functional groups bound to the polymer chain. There have been many observations supporting site isolation in suitably cross-linked functionalized polystyrenes; there have been also a number of reports of intersite interactions in polymer matrices.

Fridkin et al. made use of the possibility of bringing about an effective site isolation in resin-bound substrates, a situation approaching infinite dilution of reactants, to effect an intramolecular peptide eyclization for the synthesis of cyclic peptides (see sect. 3.4) [109,110]. Several other attempts have been made to exploit the phenomenon of diminished site-site interactions in polymeric matrices relative to those in solutions; the results are often difficult to interpret or contradictingly interpretted [137]. The titanocene derivative *19* which is bound to 20% cross-linked polystyrene shows catalytic activity, whereas its solution counter part dimerizes to the catalytically inactive *20* [138].

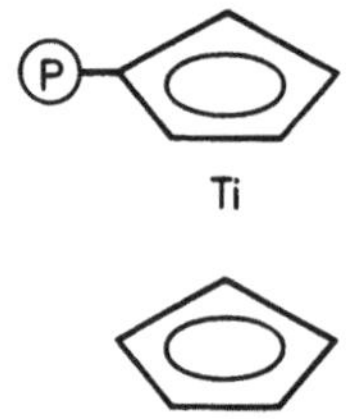


*19*

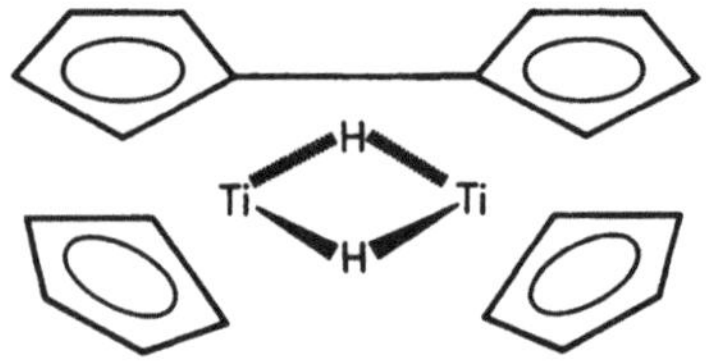


*20*

Jayalekshmy and Mazur illustrated the reduced mobility of polymer-bound reagents by delayed trapping of reactive intermediates generated within a polymer matrix [139].

Many observations, however, have provided strong evidence that site-site interactions are quite facile. Carboxylic acids bound to polystyrenes of varying degrees of cross-links have been observed to undergo anhydride formation even at low levels of functionalization [140,141]. The observation that cross-Claisen condensation products result from polymers on which two esters are attached also supports the view that intersite reactions can occur in polymer matrices [142]. Barany and Merrifield [11] analysed a number of situations under which intersite reactions occured and found that they usually occur when

1. the sites are relatively high-loaded,
2. at longer reaction times or high temperatures,

3. at appropriate distances of sites from polymer backbone, and
4. in suitable swelling solvents.

Beyerman et al. observed an intersite reaction involving the aminolysis of the peptide-polystyrene bond by the free amino group of another polystyrene-bound peptide [143]. This interchain aminolysis which occured in 2% cross-linked polystyrene resins with 0.5 to 0.7 mmole of peptide per gram of resin, in the neutralization step after deprotection, resulted in chain-doubling (scheme 4). Thus, during the neutralization of resin-bound glycine, diglycine, triglycine and tetraglycine, resin-bound diglycine, tetraglycine, hexaglycine and octaglycine respectively, were formed in small quantities (1–5%). Similarly, Rothe and coworkers during their extensive investigations on the side reactions in resin-bound peptide synthesis [144–147] observed that cross-linked polystyrene-bound proline, glycine or *N*-methylglycine produces the corresponding cyclic peptides, during neutralization, in small quantities. These cyclic peptides result by an intramolecular aminolysis of a dipeptide resin, which is formed by an intermolecular aminolysis [147,148] (scheme 5).

The possibility of achieving the infinite dilution facilitating intramolecular cyclization reactions was tried for the synthesis of a cyclopentapeptide by Rothe et al. [148]. However, they obtained the expected cyclic peptide only in negligible amount; presumably the steric isolation of the cyclizing peptide by attachment to a solid support and the exclusion of interchain reactions could not be achieved under these conditions.

Zahn and coworkers effected the interchain oxidation of resin-bound cystein peptides to cystin peptides by intersite disulphide formation in good yields [149,150].

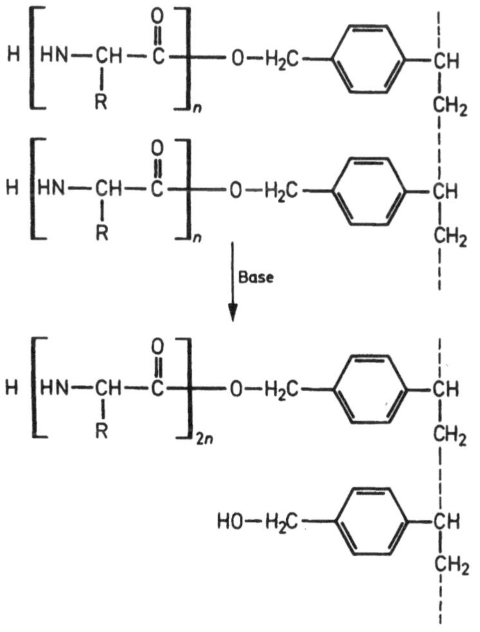


*Scheme 4.* Chain-doubling in Polystyrene-Bound Peptides.

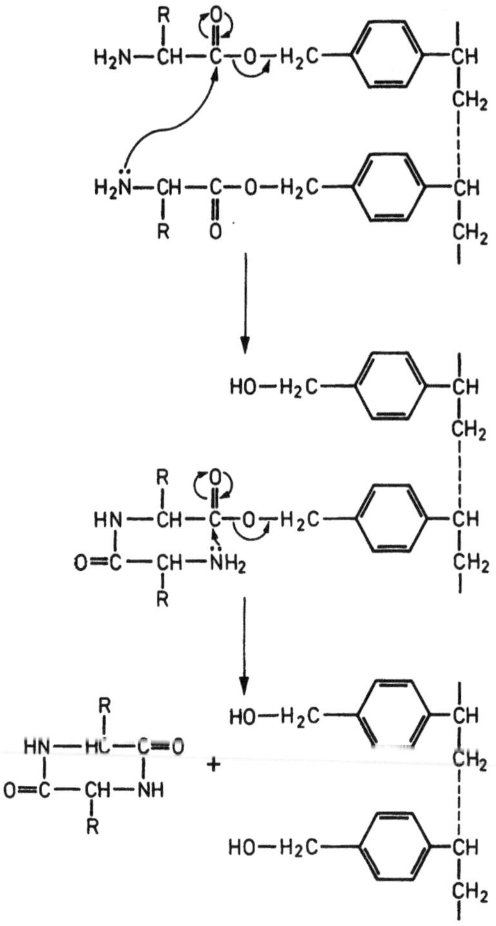

*Scheme 5.* Inter- and Intra-Molecular Aminolysis in Polystyrene-Bound Peptides.

The rate of disulphide formation in the case of the Cys-Gly-resin was found to be dependent upon various factors. The use of better resin-swelling solvents, higher loading of the bound peptide and elongation of the peptide chain increased the rate of the interchain oxidation in this case. It was also observed during these investigations that an elongation of the oxidised cystin peptide invariably results in side reactions. However, when the cystine residue was separated from the binding site on the resin by more than one amino acid residue, further elongation without side reactions was possible.

The above observations on intersite reactions and site-isolations in polymeric matrices show that by a judicious choice of the various experimental conditions, the desired effect — intersite interaction or site-site non-interaction — could be brought about. The mobility of the polymer chains depends on the degree of cross-linking, solvent, temperature, flexibility of the chain segments and the distribution of the substituents. High cross-link density, non-swellability, low temperature etc. tend the polymer to behave as a rigid matrix [15]. However, the reactivity of the system under these conditions is minimal and effective penetration of reagents occurs in swollen systems only.

## 4.4 Solvation and Swellability of Peptide Resins

In cross-linked polymeric supports, the extent of swelling in a particular solvent is determined mainly by the degree of cross-linking. The degree of cross-linking also determines the chemical reactivity of functional groups, the effective pore-size and the physical stability of the macromolecular beads. A comparison of these properties for lightly and heavily cross-linked polystyrene resins shows that beads containing 1–2 percent divinylbenzene are still compatible with good physical stability and at the same time swell good in organic solvents. In the synthesis of Ribonuclease A, Gutte and Merrifield observed that with the growing chain length of the peptide, the volume of the 1% cross-linked polystyrene resin increased considerably, but its swelling and filterability behaviour as well as its mechanical properties are not noticeably affected [151].

For the effective coupling of amino acid residues to the polymer-bound peptide chain, the reaction should be carried out in solvents which swell the resin. Wieland et al. determined the swelling factors for a 2% cross-linked polystyrene-bound Boc-Phe resin in different solvents [152]. The volumes of 1 g of this Boc-Phe resin after treatment for 50 hours in different solvents are given in Table 1. These values show that in the case of the polystyrene resin, swelling is minimum in very polar solvents (like methanol) or in very nonpolar solvents (like hexane) and is greatest in chlorinated solvents (like methylenechloride).

In the case of the polar polymeric support, poly(*N*-acrylylpyrrolidine) (PAP) resin, Walter and coworkers observed favourable swelling properties for the protonated or acylated form in a wider range of solvents ranging in polarity from water to methylenechloride [65]. Comparison of the swelling properties of Boc-Gly-PAP resin and Boc-Gly-polystyrene resin showed that the former is superior to the latter in its

**Table 1.** Swelling of Crosslinked (2%) Polystyrene-bound Boc-Phe Resins

| Solvent | Volume of 1 g of bead[a] after swelling for 50 hours, ml |
|---|---|
| Methylenechloride | 6.2 |
| Chloroform | 5.7 |
| Tetrahydrofuran | 5.3 |
| Dioxan | 5.2 |
| Dimethylformamide | 5.1 |
| Ethylacetate | 4.7 |
| Acetic acid | 2.9 |
| Ethanol | 2.2 |
| Water | 1.5 |

*a* dry volume 1.6 ml

**Table 2.** Swelling Properties of Crosslinked Poly(*N*-acrylylpyrrolidine)[a] and polystyrene[b] resins

| Solvent | Volume of 1 g of bead after equilibration in the solvent | | |
|---|---|---|---|
| | PAP-HCl | Boc-Gly-PAP[c] | Boc-Gly-PS[d] |
| Acetone | 1.5 | 2.4 | 3.8 |
| Methanol | 9.5 | 6.6 | 1.6 |
| Ethanol | 8.1 | 7.3 | 1.8 |
| Isopropanol | 7.3 | 6.5 | 2.1 |
| Dimethylformamide | 6.4 | 5.2 | 5.2 |
| *N*-Methylpyrrolidinone | 5.4 | 6.2 | 6.5 |
| Methylenechloride | 6.4 | 7.5 | 7.6 |
| Chloroform | 6.8 | 8.7 | 9.5 |
| Ethylacetate | 1.9 | 1.9 | 4.7 |
| Acetic acid | 7.9 | 8.5 | 2.5 |
| Water | 10.5 | 6.2 | 1.6 |
| Trifluoroethanol | 14.1 | 11.2 | 1.6 |

*a* Degree of cross-linking, 4.4; Monomer Dilution Ratio, 4
*b* Degree of cross-linking, 1
*c* 0.58 mmol of Gly per gram of resin
*d* 0.57 mmol of Gly per gram of resin

ability to interact with polar solvents like alcohol and water (Table 2). The PAP resins also swell good in methylenechloride, chloroform and dimethylformamide, which are the commonly used solvents for the polystyrene resins.

The above observations on the swelling properties of the resin supports in different solvents point to the fact that the extent of swelling and solvation is a very decisive factor for the successful execution of the solid phase synthesis of longer peptides [153].

Sarin, Kent and Merrifield carried out a very interesting investigation of the factors affecting the solvation and swelling of 1% cross-linked polystyrene-bound peptides [154]. They found that the original volume of 1 g of the unsubstituted dry beads increasing by more than fivefold as the peptide content approaches 80%. At the same time, the swollen volume of the resin in methylenechloride increased from 6.2 ml to 12 ml; in DMF the values were 3.3 ml and 28 ml. These studies indicated no upper limits of swelling of the peptide resins. The solvation properties of the cross-linked polymer network and the attached peptide chains mutually affect one another. At high loadings, the peptide component has a dominating influence on the swelling of the resin-peptide in a solvent such as DMF. Sarin et al. have also offered an explanation in thermodynamic terms for the increased swelling of such resin-peptides. The swelling of the unsubstituted resin is due to a drop in free energy resulting from the solvation of the polystyrene matrix, and at equilibrium, this is balanced by the elastic restraining force resulting from the deformation of the loosely cross-linked polymer network. Thus the increased swelling of the peptide resin is due to the additional net decrease in free energy from solvation of the linear peptide chains; this is counteracted by an increase in the elastic restraining force arising from further deformation of the loosely cross-linked polymeric network.

No such additional counterforce is expected to arise from the deformation of the linear peptide chain imposed by the expansion of the polymeric support.

The network structure of the cross-linked polymer minimizes the self aggregation of the peptide and polystyrene components [154]. This leads to a structure for the amorphous peptide-resin in which the two dissimilar components are intimately mixed at a molecular level, and which is less favourable thermodynamically [155] than is the case for self aggregated free components. Both the polymer and the peptide will tend to interact more favourably with the added solvent and will exert a mutual solubilizing effect on one another. Thus, solvents which normally would be unable to dissolve a protected peptide can be effective solvating agents for the same resin-bound peptides in organic solvents. These observations on the swelling and solubilizing behaviour of loosely cross-linked polystyrene resins are of very much significance in the exploitation of the solid-phase method for the synthesis of larger peptides and proteins.

Birr [156] finds that the reactions on supports like 1% cross-linked polystyrene in suitable solvents are not strictly "solid-phase" reactions, even if the particles look solid. Under these loosely cross-linked conditions, the polymer exists in a quasi-dissolved gel state, and the reactivity of the chain molecule is not considerably reduced in any way by the gel character. The free motion of the large chain segments in the swollen state of the macromolecular gel phase — which is only very lightly cross-linked just to give macroscopic insolubility — guarantees reactivity which is not much different from the corresponding linear, soluble macromolecule.

## 5 Techniques with Soluble Polymers

A number of linear, non cross-linked soluble polymers have been investigated for their use as supports for biopolymer synthesis and as reagent carriers in organic synthesis [157]. The motivation behind all these attempts have been mainly the circumvention of the diffusion and reactivity problems often encountered in the heterogeneous solid phase reactions.

Thus, for example, Ovchinnikov and coworkers [158,159] tried to eliminate the steric complications encountered in the original solid phase method by employing linear, non-crosslinked polystyrene as a soluble support for stepwise peptide synthesis. The coupling and deprotection reactions were carried in solution and the excess reagents were removed by precipitation of the polymer-bound peptide in water and washing. This synthesis in homogeneous phase was claimed to be superior to the heterogeneous reaction on account of the higher chances of achieving quantitative coupling reactions.

However, Green and Garson [160] observed that the use of soluble polystyrene involves the danger of cross-linking during the various synthetic steps. This is mainly due to the residual chloromethyl groups remaining after the esterification step. However, when special care was taken to avoid cross-linking reactions, distinct advantages with respect to the elimination of steric effects of the support could be demonstrated using the linear polystyrene [160]. Maher et al. have also used soluble linear polystyrene for peptide synthesis [122].

Andreatta and Rink [119] synthesized model-peptides on a linear polystyrene of molecular weight 20,000, employing the gel filtration technique for the separation of the polystyrene-bound peptides from excess reagents. These investigators observed that the maximum leading to maintain satisfactory solubility properties of the polymer-peptide was 0.5 mmol peptide per gram of polymer.

Water-soluble polyethylenimine of molecular weight 30,000 has been used for the synthesis of a model tetrapeptide by Blecher and Pfaender [161]. The *N*-carboxyanhydride method was used for the coupling, and by ultrafiltration excess reagents were removed. An enzymatic method was employed in this case for the splitting of the peptide from the support.

Bayer and Geckeler used copolymers of *N*-vinylpyrrolidinone and vinylacetate for the synthesis of peptides [162,163]. With symmetric anhydrides as coupling agents, several peptides could be synthesized on these supports in overall yields of *ca.* 60%. Owing to the presence of the pyrrolidinone ring, these copolymers exert a strong solubilizing effect upon the peptide chain in a variety of solvents including water. However, the same syntheses using poly(acrylic acid) and poly(vinylalcohol) as supports gave less satisfactory results, mainly due to the poor solubility of the polymer-bound analogues. However, none of these linear soluble polymers has found general applicability in peptide synthesis.

Kinetic investigations have been carried out in order to delineate the differences in the reaction rates between low and high molecular weight peptide esters [119]. A result of major impact for the use of the above linear polyfunctional supports, in general, was the finding that the functional groups attached to polystyrene showed considerable differences in chemical reactivity and kinetic behaviour. Thus, for example, in the hydrogenolytic cleavage of the benzyl ester groups joining the peptide with the polymer support, some anchoring groups proved to be totally resistant. The kinetic course of the coupling reaction also showed significant deviations from linearity due to the non-equivalence of functional groups. The enhanced reaction rates of some specific groups appear to originate from favourable polymer effects such as an increase in the effective dielectric constant at the reactions site [164]. A comparison of the second order rate constants for the aminolysis of N-protected active esters with low- and high-molecular weight amino components shows that the reaction rates for the soluble polymer ester are lowered by a factor of about two compared to low molecular weight esters, but considerably higher than those of the heterogeneous system (Table 3).

The above-observed differences in the chemical reactivity of the attached functional groups in the linear polymeric supports hinder the quantitative coupling under normal conditions, even though the formation of truncated sequences due to sterically in-

**Table 3.** Second Order Rate Constants for the Aminolysis of Z—Ala—ONp with H—Pro—O—$CH_2$—R

| R | $K_2$, liter · mole$^{-1}$ · sec$^{-1}$ |
|---|---|
| Phenyl | 0.081 |
| Linear polystyrene | 0.047 |
| 2% Crosslinked polystyrene | 0.030 |

accessible reaction sites is unlikely in the homogeneous system. A possible reason for the heterogeneous kinetic behaviour of the linear polyfunctional supports is the random distribution of functional sites along the linear chain molecule. Thus, functional groups located at the chain ends might be less subject to local effects of the polymer chain, than groups near the centre of the chain, due to the differences in the segment flexibility and in the density of the coiled macromolecular chain.

These results point to the fact that the steric complications inherent of the heterogeneous solid-phase reactions occuring in cross-linked polymeric matrices are not solved completely by the use of the corresponding non-crosslinked soluble polymeric supports. The equivalence of all functional groups attached to the linear macromolecular chain, therefore, appears to be a prerequisite for the attainment of the reaction facility prevailing in low-molecular weight systems.

## 5.1 Polyethers as Soluble Supports

The above drawbacks of the linear polyfunctional macromolecular supports are to a greater extent overcome by the use of appropriate polyethers as soluble supports for biopolymer synthesis. Absence of any steric effect, equivalence of functional groups and compatibility with the biopolymers being synthesized can be expected from polyethyleneglycols, which are linear polyethers with two hydroxy groups at the chain ends (*21*).

$$HO\text{-}(CH_2-CH_2-O)_n H$$

*21*

The presence of both a hydrophilic and hydrophobic moiety per monomer unit lends this class of polymers its favourable physical and chemical properties for peptide synthesis. From a theoretical point of view, the reactivity of the two hydroxy groups in polyethers, such as polyethyleneglycols, should be identical and of the same order as that of the corresponding low-molecular weight analogues [165].

A comparison of the rate constants of the aminolysis of N-protected amino acid esters with both amino acid PEG esters of different molecular weights and with their low-molecular weight analogues shows no significant difference in reactivity [166]. Also, no significant influence of the molecular weight of the polyethyleneglycols could be detected in the range of molecular weight 2,000–20,000 (Table 4). Within this range, the reaction between the active amino acid esters and the amino acid PEG esters strictly followed second order kinetics, from the beginning of the coupling until about 80% completion. Thus, PEG esters showed fully identical behaviour to the low molecular weight esters, and no unfavourable influence of the peptide chain upon the reaction rates could be detected [167]. However, the fact that the reaction rates for the low-molecular weight esters and PEG esters are identical is not an evidence for the linearity of the kinetics up to 100% completion of reaction. Although the final stage of the reaction is difficult to assess in these kinetic investigations, potential deviations from linearity due to sterically heterogeneous functional groups, as observed in the case of cross-linked polymeric supports, can be excluded here for several reasons: the functional groups in PEG are equivalent and are con-

**Table 4.** Rate Constants for the Aminolysis of Boc—Gly—ONp with H-(Gly)$_n$-OR

| R | n | $K_2$, liter · mole$^{-1}$ sec$^{-1}$ |
|---|---|---|
| Ethyl | 1 | 0.013 |
| Ethyl | 3 | 0.019 |
| *t*-Butyl | 1 | 0.024 |
| $-CH_2-CH_2-OCH_3$ | 1 | 0.008 |
| PEG 20,000 | 3 | 0.032 |
| PEG 20,000 | 1 | 0.014 |
| PEG 10,000 | 1 | 0.014 |
| PEG 6,000 | 1 | 0.016 |
| PEG 4,000 | 1 | 0.019 |
| PEG 2,000 | 1 | 0.018 |

sequently in identical physicochemical environment; in view of the peculiar conformational nature of PEG, local steric effects are unlikely because of the high flexibility of the chain, and diffusion effects are also absent in homogeneous solution even when the viscosity is relatively high [164].

Investigations using polyethyleneglycols as soluble supports for peptide synthesis confirm the above expectations. Thus, for example, in hydrogenation reactions, which revealed considerable heterogeneity of functional groups in non-crosslinked polystyrene supports [119], no influence of the polyethyleneglycol support on the reaction rate could be detected and the reaction proceeded to 100% completion [168]. Mainly polyethyleneglycols of molecular weights ranging from 2,000 to 20,000 and monofunctional polyethyleneglycols have been used extensively as supports for peptide synthesis [169]. In order to obtain higher polyether capacities, block copolymers with functional groups in definite distances were synthesized starting from polyethyleneglycol blocks and diisocyanate derivatives by polyaddition reactions as illustrated in scheme 6 [170]. When PEG of molecular weight 1,000 was used as the starting material for this polyaddition reaction, block-copolymers of molecular weight 20,000 were

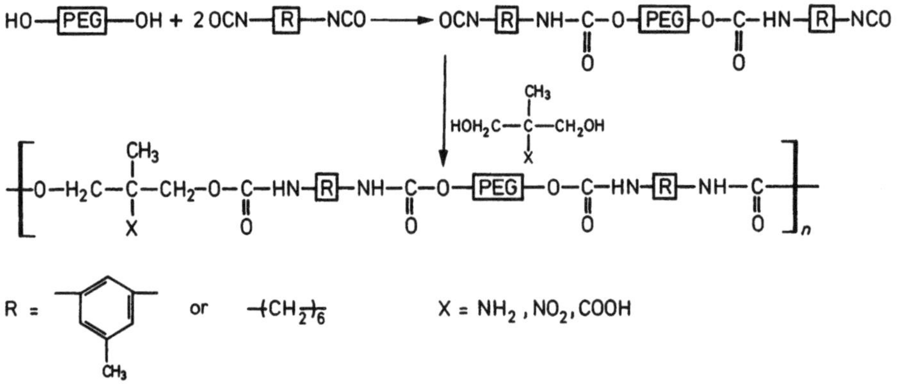


*Scheme 6.* Synthesis of Block Copolyethers with Regularly Placed Functional Groups.

obtained. The functional groups are so located at definite distances in these block-copolyethers that neighbouring group effects are unlikely [171].

## 5.2 Peptide Synthesis on Polyethyleneglycol Supports

The synthesis of peptides on polyethyleneglycol supports — Liquid Phase Method (LPM) — takes advantage of the concept of stepwise synthesis on a passive macromolecular carrier, without intermediate isolation and purification procedure [172,173]. In this method, the C-terminal amino acid is fixed to the solubilizing polyethyleneglycol, which determines the physicochemical properties of the growing peptide chain during all stages of synthesis. As a result, the polymer-bound peptide can be readily separated from low-molecular weight reagents by selective precipitation, and the stepwise incorporation of the amino acid residues is as efficient as in the solid phase method. All the coupling and deprotection reactions are carried out in homogeneous solution as in the classical peptide synthetic procedures, thereby eliminating the intrinsic difficulties of the heterogeneous reactions. Moreover, compared to the classical stepwise strategies, the use of the PEG support, which acts as a solubilizing C-terminal macromolecular protecting group, offers two significant advantages:

1. The solubility of the peptide is strongly enhanced by the polymeric ester group so that peptides with poor solubility become accessible to the stepwise strategy.
2. The synthesis cycle is simplified and performed according to a standard procedure independent of the physicochemical properties of the peptide.

For an efficient synthesis using the liquid phase method, the crystallization tendency and the solubility of the polyethyleneglycol in different organic solvents must be retained after the attachment of the peptide to its chain ends. The influence of the attached peptides on these properties of the polymer was found to depend on the primary sequence, side chain protection, chain length and conformational preferences of the growing peptide chain. X-ray investigations on polyethyleneglycols of varying molecular weights and on different PEG-bound peptides showed that the incorporation of the peptide does not disturb the crystal lattice of PEG, and its degree of crystallization is lowered only by a relatively small factor [174,175].

This high retention of the crystallinity of PEG after the attachment of the amorphous peptide blocks can be understood when we consider the conformation of pure polyethyleneglycols in the solid state. X-ray studies reveal that chain-folding occurs in crystals of PEG when the molecular weight is greater than 3,000 [176-178].

Investigations on a block-copolymer of PEG and polystyrene also showed that in the solid state, the amorphous polystyrene occupies a space between two PEG blocks [179]. These observations suggested a similar two-phase model for the PEG-peptides in the solid state which explains the high retention of crystallinity of the polymer even when it is bound to amorphous peptides [180].

According to this model, crystalline PEG layers alternate with the amorphous peptide phase; due to the regular folding of the PEG chains, the bulky peptide coils are arranged between the PEG blocks without disturbing the crystal structure of PEG. This finding is of practical relevance for the liquid phase peptide synthesis because the partially crystalline structure of the peptide-PEG esters permits easy handling of the solid precipitates. Moreover, the danger of inclusion of low molecular weight components, as observed in the amorphous precipitates in the case of polystyrene-bound peptides [158], is considerably reduced. Thus, a single precipitation of the polyethyleneglycol-bound peptides is often sufficient to remove the excess of soluble reagents after the coupling and deprotection reactions. Even at equimolecular proportions of the peptide and PEG, the polymer-peptide could be precipitated quantitatively from organic solvents and showed partially crystalline structure.

The solubility of the polymer-bound peptide is also of utmost importance in the application of the liquid phase method to repetitive sequential-type synthesis. The solubility of the PEG-peptides is generally similar to that of the unbound polymer. However, this property is highly dependent on the conformation of the particular peptide sequences. Thus, the solubility of the PEG-esters of the hydrophobic homo-oligomers $(Ala)_n$, $(Val)_n$ and $(Ile)_n$ decreased considerably for chain lengths up to 10 residues by the stepwise liquid phase method. Conformational investigations reveal that the low solubility of these homooligopeptides result from their tendency to form aggregated β-structures with intermolecular hydrogen bonds [181,182]. In some cases the solubility increases on further elongation of the PEG-peptide. This behaviour is generally the result of a conformational transition from aggregated β-structures to helical structures.

Similar observations have been made during the stepwise synthesis of other sequential peptides [174,183]. These results indicate the impact of conformational transitions in the synthetic strategies of peptide sequences (see Sect. 5.6). Variation of the molecular weight of PEG in the range of 2,000 to 20,000 showed no dramatic effect on the solubility of the attached peptides. Molecular weights higher than 20,000 have very low capacity and the lower molecular weight polyethyleneglycols are no longer amenable to crystallization. However, suitable derivatives of low molecular weight polyethyleneglycols find application as solubilizing protecting groups for the synthesis of peptides [184,185].

## 5.3 The Technique of the Liquid Phase Method

Typically, the liquid phase method of peptide synthesis combines the strategical features of the classical and solid phase methods. Through the presence of a macromolecular protecting group, PEG, at the C-terminus of the peptide chain, the most important advantage of the polymer-supported synthesis — efficient stepwise synthesis without intermediate purification procedures — is preserved; at the same time, quantitative coupling and deprotection are facilitated by working in a homogeneous solution. The polyethyleneglycol-bound peptides can be purified by selective crystallization from organic solvents. A general picture of the various steps involved in the liquid phase method is outlined in scheme 7.

### 5.3.1 Attachment of the First Amino Acid and Anchoring Groups

In the liquid phase method, the C-terminal N-protected amino acid of the desired peptide sequence is esterified to the terminal hydroxyl group of PEG [174]. By attaching a suitable anchoring group to the chain ends of PEG, prior to the attachment of the first amino acid, the chemical stability of the polymer-peptide bond

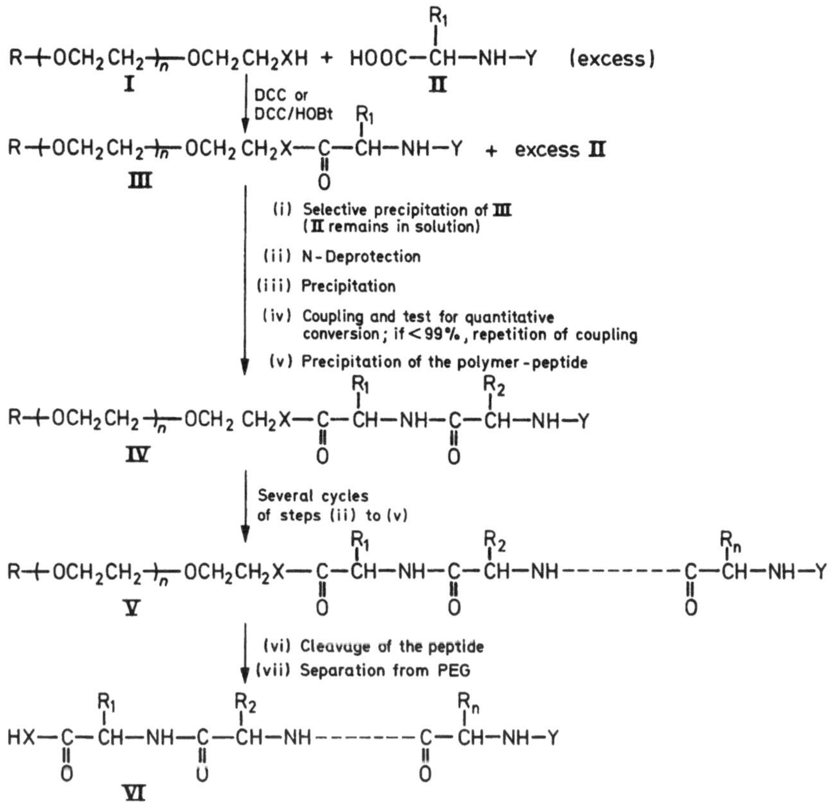


*Scheme 7.* The Liquid Phase Peptide Synthesis. (In this scheme R = $CH_3$ or $CH_2CH_2XH$; X = 0, NH or their combination with anchoring groups).

can be modified. PEG derivatives obtained by the replacement of the terminal hydroxyl groups by various other groups like amino halogeno etc. could also be made use of for the liquid phase method [186-188]. As the stability of the bond between the polyoxyethylene chain and the C-terminal amino acid of the peptide sets limits upon the choice of the side-chain protecting groups, tailoring of the anchoring groups should allow greater flexibility in the concerned synthetic approaches. The direct esterfication of PEG with the N-protected amino acid results in an ester bond with chemical stability similar to that of the low molecular weight aliphatic esters. This ester bond could be hydrolyzed under mild alkaline reaction conditions.

However, because of the danger of racemization and other side reactions involved in the alkaline hydrolysis, the use of an aliphatic ester bond — direct attachment to PEG without the use of an anchoring group — is not desirable in many cases. Attempts to cleave this aliphatic ester bond under acidic reaction conditions were not successful. On the other hand, the peptide derivatives were cleaved from the polyethyleneglycol supports by transesterification, hydrazinolysis or aminolysis [174].

The benzyl ester link which is predominantly used in the solid phase method, have been used in the PEG-based synthesis; it permits mild acidolytic, basic or catalytic hydrogenolytic cleavage. Monofunctional PEG containing benzyl bromide functional group has been used as a support for the liquid phase synthesis of a decapeptide corresponding to the last 10 amino acid residues of bovine insulin B-chain [189]; cleavage of the fully protected peptide from the polymer was effected with 1N NaOH in dioxane. Replacement of the terminal hydroxyl groups of PEG with amino groups permitted a more facile introduction of the anchoring groups. Reaction of PEG, as well as its terminal amino analogue, with 4-(bromomethyl)-3-nitrobenzoic acid resulted in photolytically removable soluble polymeric supports for the stepwise synthesis of fully protected peptides [190–193].

This 2-nitrobenzyl anchoring group, introduced by Rich and Gurwara in the solid phase peptide synthesis [194–196], is stable under all conditions of the LPM and permits almost quantitative release of peptides by irradiation at 350 nm without affecting aromatic amino acids like Tyr, Phe or Trp (Scheme 8). Such a support has recently been used for the synthesis of the fully protected C-terminal 21-peptide corresponding to the insulin A chain [197].

4-Aminomethyl-3-nitrobenzoylpolyethyleneglycols are another class of photosensitive soluble polymeric supports designed for the liquid phase synthesis of protected peptide amides [192,193] (scheme 8). These supports have been very recently used for efficient synthesis of three biologically active 14-peptideamides corresponding to the wasp venom peptides, mastoparan, mastoparan X and Polistes mastoparan [198].

Monofunctional polyethyleneglycol supports containing benzyloxycarbonylhydrazide, *p*-benzyloxybenzyloxycarbonylhydrazide and *t*-butyloxycarbonylhydrazide anchoring groups, *22*, *23* and *24* respectively have been developed recently for the liquid phase synthesis of protected peptide hydrazides [199].

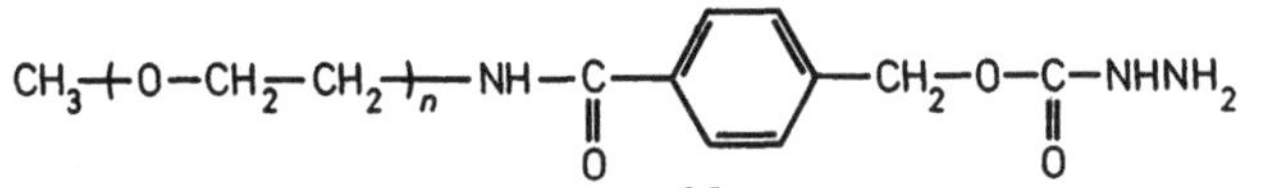

*22*

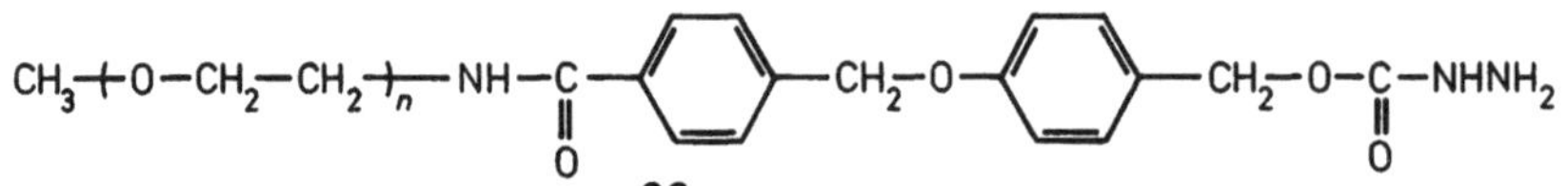

*23*

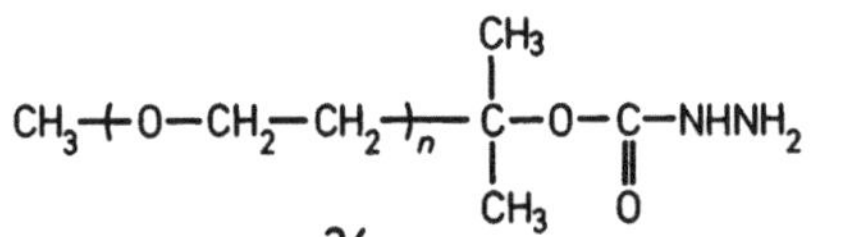

*24*

The anchoring bond between the peptide chain and the polymer *22* permits cleavage by catalytic hydrogenolysis, thus leaving other *t*-butyl based protecting groups unaffected. Bonds formed by supports *23* and *24* are acid-labile and can be selectively cleaved by 50% trifluoroacetic acid in methylenechloride; under these conditions the benzyl based protecting groups and Fmoc group are stable, thus allowing the preparation of fully protected peptide hydrazides. The PEG support *25* with *p*-alkoxy-

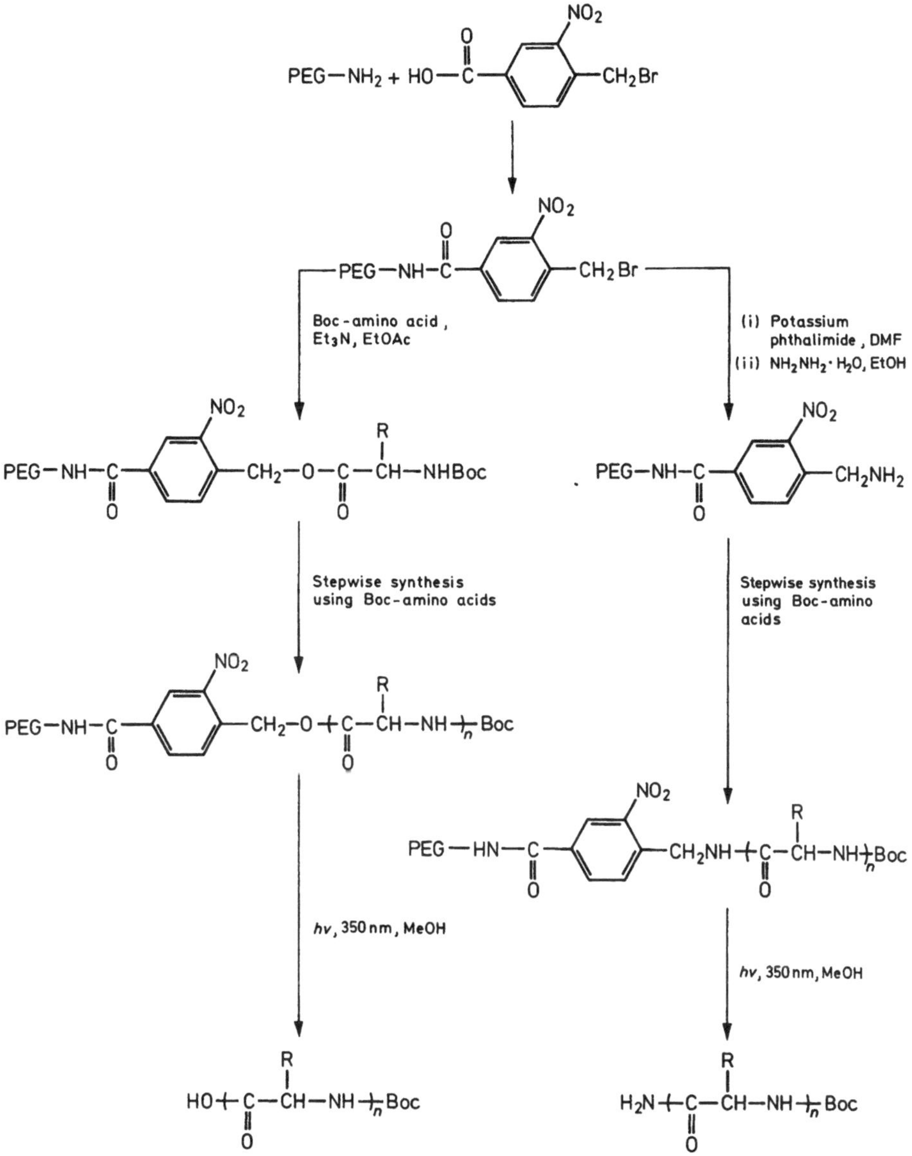


*Scheme 8.* Photolabile Anchoring Groups for the Liquid Phase Peptide Synthesis.

benzyl anchoring group has been prepared by Colombo and Pinelli [200]; this group permitted cleavage of the finished peptides with trifluoroacetic acid. The liquid phase synthesis of two pentapeptides corresponding to the amino acid sequences of Leu- and Met-enkephalin, using the support *25* has been described [200].

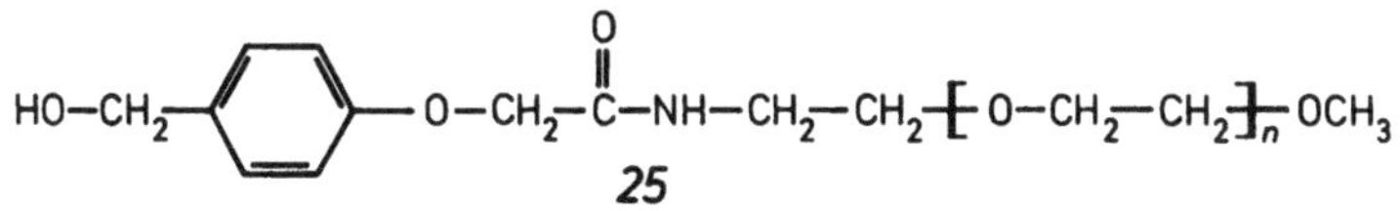


*25*

### 5.3.2 Coupling and Deprotection Steps

The coupling reaction in the LPM is usually carried out in methylenechloride; solvents like DMF and DMSO are also used in the later stages where solubility in $CH_2Cl_2$ is limited. In the coupling processes, the neutralization of the protonated terminal amino groups of the peptide-polymer by a tertiary base is performed subsequent to the addition of the activated carboxyl component in order to avoid a high local concentration of the amino groups at any time. This is particularly important at the dipeptide stage, because of the possibility of the diketopiperazine formation. In cases where long reaction times are necessary to achieve quantitative coupling, the presence of salts may give rise to some degree of racemization. Neutralization and removal of salts precipitation of the N-deprotonated PEG-peptide before the addition of the activated carboxyl component might be advantageous in these instances.

For the formation of the peptide bond, all the commonly used methods in the classical synthesis can be made use of. The use of symmetrical anhydrides, prepared separately by reacting two moles of N-protected amino acid with one mole of dicyclohexylcarbodiimide, proved to be convenient, since insoluble dicyclohexylurea can be removed prior to coupling [201]. DCC activation in presence of hydroxybenzotriazole has also been observed to be equally effective for the coupling reaction in the liquid phase method [198,202]. Usually, 2- to 3-fold excess of the carboxylic component is used. Depending upon the molecular weight of PEG, the coupling reaction is carried out in $10^{-1}$ to $10^{-2}$ molar solution, which is about 10 times higher than in solid phase peptide synthesis. A gradual increase in concentration of the solution by continuous evaporation of the solvent during the coupling reaction helps to maintain reasonable reaction rates in the final stages of the coupling.

After the coupling reaction, the peptide-PEG ester is precipitated by adding diethylether to the reaction mixture. Excess reagents remain in solution and the polymer-bound peptide is removed by filtration. Because of the relatively high crystallinity of the peptide-PEG esters, they are obtained as granular precipitates and can be easily filtered and dried. Depending upon the solubility of the N-protected carboxyl component, the isolation and purification of the polymer-bound peptide are accomplished by selective precipitation with ether and/or crystallization from ethanolic or methanolic solution. In extreme cases, excess reagents in homogeneous solution can be removed by ultrafiltration, solvent extraction or chromatographic procedures [174]. Before the deprotection of the terminal amino group, the coupling yield is determined quantitatively. In the case of incomplete reaction, the coupling step is repeated under identical or under more favourable conditions. The reaction period for obtaining quantitative coupling depends on factors such as steric contributions from the coupling components, solvation and conformation of the peptide chain. Thus, the optimum coupling conditions in terms of the method of activation, excess reagents, solvent and reaction period are mainly sequence-dependent. The use of competitive experiments [203–205] in the case of the polyethyleneglycol-bound peptides provides valuable information about these factors and is helpful in deciding for a suitable combination of the variables of the coupling reaction conditions in the liquid phase method [206].

The temporary amino protecting group most commonly employed in the liquid phase method is *N-tert*-butyloxycarbonyl (Boc). This group can be removed from the PEG-peptide by treatment with trifluoroacetic acid in methylenechloride (1:1) or by 1.2 N HCl/HOAc. When HCl in organic solvent is used for deprotection, the concentration of HCl must be increased by about 10% in order to compensate for the formation of the oxonium salts of PEG. The N-deprotected peptide-PEG ester is isolated as usual by the addition of ether to the reaction mixture after removal of the acids. In order to remove the last traces of acid, the polymer-peptide is reprecipitated from ethanolic solution. The benzyloxycarbonyl group when used in the LPM can be deprotected by catalytic hydrogenation as in the classical procedure. The completion of the deprotection reaction can be ascertained by potentiometric titration of the free amino groups or by IR and NMR spectral analysis.

### 5.3.3 Analysis and Automation

The solubilizing effect of PEG on the attached peptide and the absence of any direct influence of the polyoxyethylene chain on the physicochemical properties of the peptides provide a wider range of possibilities for analytical control during the liquid phase peptide synthesis than those in the solid phase method. The reactions employed in the stepwise liquid phase synthesis can thus be quantitatively monitored by several analytical methods.

For the analytical control of the coupling reaction, the most predominantly employed techniques are the fluorescamine and ninhydrin methods [207]. For a rapid quantitative assessment of the extent of coupling, an aliquot of the reaction mixture containing about 20 µmol of the peptide-PEG ester is precipitated with ether, filtered and treated with the ninhydrin reagent; the colour developed in this case is compared with that of a sample of 0.5 µmol of uncoupled peptide-PEG ester developed analogously, and the coupling yield is estimated roughly as 98% when the intensity of colour is the same in both cases. Unreacted amino groups even to the extent of less than 0.1% can be detected by the ninhydrin method when a scanner or amino acid analyser is used for quantitative evaluation. The fluorescamine method and IR spectroscopy in conjunction with stop-flow technique, has been used for monitoring the kinetic course of the coupling reaction [207,208].

NMR techniques have been investigated for the analytical control of the liquid-phase peptide synthesis. The method offers a nondestructive analytical technique to directly monitor the progress of the peptide synthesis by characterizing the peptide without cleavage from the polymeric support or disrupting the synthetic steps. Leibfritz et al. investigated the applicability of the $^{13}$C-NMR methods for the analytical control of the stepwise synthesis of peptides on a PEG support [209]. However, large amounts of the polymer-peptide and long overnight accumulations were necessary to obtain well-resolved $^{13}$C-NMR spectra at natural abundance. Very recently, Ribeiro et al. have demonstrated high resolution 360 MHz $^{1}$H-NMR to be an elegant method for the rapid characterization of synthetic peptides attached to PEG in the liquid phase method [210]. A combination of the high field and selective saturation or Redfield pulse sequence method [211] in this case allows resolution of individual backbone NH and α-CH resonances of dilute peptides in the presence of the strong resonance from PEG and/or protonated solvents.

The quantitative deprotection of the terminal amino group can also be ascertained by potentiometic microtitration, ninhydrin or fluorescamine colorimetric methods and by spectroscopic techniques such as IR or NMR. For the amino acid analysis in the case of the peptide-PEG esters, the sample is subjected to hydrolysis and analyzed directly; the soluble polymer is eluted at the front. Similarly, the total number of amino groups can be determined without cleaving the PEG support from the peptide by subjecting the terminal amino peptide-PEG ester directly to amino acid analysis.

The automation of the main operations in the synthesis cycle of the liquid phase strategy has been realized [205]. A total programmed cycle for the incorporation of one amino acid residue in this synthesizer requires a period of 2 to 5 hours depending upon the method of coupling and the amino acid. Time-programming on the basis of predetermined coupling periods was used for this automatic procedure for the liquid phase synthesis. Rapid synthesis, elimination of operative errors and increased yields are the major advantages of this automatic procedure. Owing to the equivalence of the functional groups in PEG, high reproducibility is guaranteed in the standardized procedure.

However, the change in the physicochemical properties of the polymer-bound peptide with increasing chain lengths represents a potential limitation for the application of the standard procedures for the synthesis of larger peptides. The attempts to design a liquid-phase peptide synthesizer making use of membrane-filtration instead of crystallization and filtration (for reducing the number of steps involved) in order to remove excess low molecular weight reagents were unsuccessful; this is due to the varying membrane-permeability of the individual polymer-peptides [205].

Automation of the liquid-phase synthesis, at its present stage, is advantageous only when the optimum reaction conditions for the synthesis of a specific peptide have been elucidated in advance, and the synthesis of small- to medium-sized peptides appears to be within the scope of this automated method.

## 5.4 Solid-Liquid Phase Techniques

The strategical features of the liquid phase, solid phase and the polymer reagent methods are combined in the so-called alternating liquid-solid phase method for peptide synthesis introduced by Frank et al. [212,213]. In this method, the activated carboxyl component is attached to an insoluble polymer matrix by an acid-labile urethane bond. This facilitates the removal of the unreacted components from the main product after coupling by a simple filtration procedure. The temporary immobilization of the growing peptide chain during the coupling reaction eliminates the absolute need for quantitative reactions without losing the practical advantages of the solid phase method. A large excess of the polymeric active ester can be used in order to facilitate effective coupling. The removal of the excess coupling components and reagents is facilitated by the use of a soluble polymeric carboxyl protecting groups as in the liquid phase method. In a typical case, a PEG-bound dipeptide with deprotected terminal amino group is coupled with the resin-bound active ester of the N-protected amino acid to give the extended peptide-PEG ester coupled to the resin. This is followed by acidic cleavage to give the N-terminal extended peptide PEG-ester (scheme 9). This cycle involving the two steps — coupling and deprotection — is repeated to produce the desired peptide.

In this alternating liquid-solid phase peptide synthesis, unlike in the solid phase

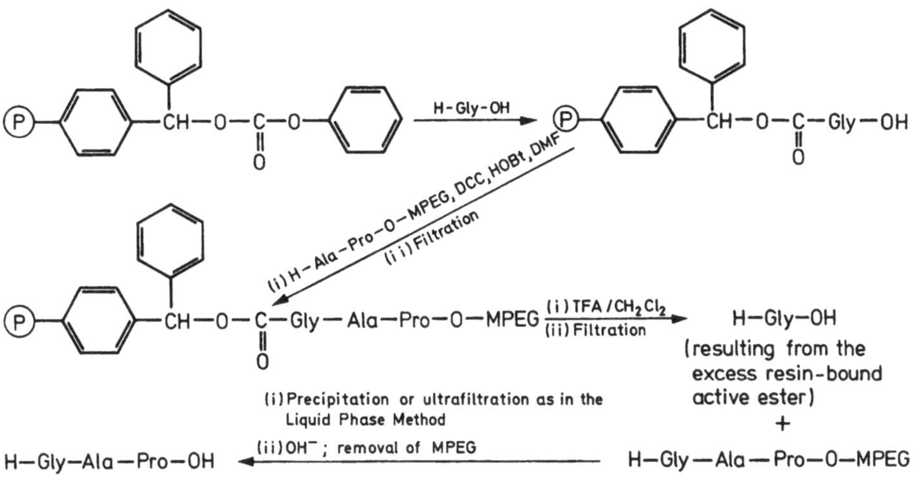


*Scheme 9.* Alternating Liquid-Solid Phase Method.

method, the incomplete coupling and/or deprotection reactions do not lead to truncated or deletion peptides, since each byproduct can be removed before proceeding to the next synthetic step [214]; however, such incomplete reactions reduce the yield of the elongated peptide. The high purity of the final product can be expected from this strategy and this has been illustrated in the synthesis of LH—RH [213]. The limiting factors of this method are the necessity of a high excess of resin-bound amino acid derivatives and the limitation of the soluble component to a molecular weight of about 4,000 [214,215].

A combination of the liquid-phase method and the polymer reagent technique has been applied for the synthesis of peptides by Jung et al. [216,217]. In this method the soluble, polyethyleneglycol esters of the C-terminal amino acids are treated with excess of active esters covalently bound to insoluble, cross-linked polystyrene beads. After the completion of the coupling step, the excess reagent is removed by a simple filtration step instead of precipitating the PEG-peptide as in the liquid phase method (scheme 10).

Thus, in this method the stepwise synthesis proceeds on the soluble polymer, as in the liquid phase method; the only difference is in the coupling step where the carboxyl component is an insoluble polymeric active ester. The use of the solubilizing C-terminal macromolecular protecting group overcomes the limitations encountered in the original application of the polymer reagent method of peptide synthesis. However, due to the steric limitations characteristic of the polymer-polymer interactions, the attainment of quantitative coupling in each step of the synthesis can be difficult in this method.

The possibility of providing an orthogonal [91,92] set of anchoring linkages for binding the solid support, soluble support and the substrate among one another in order to design reversibly immobilized supports for biological macromolecules is at present under investigation in our laboratory [218]. A typical system in this case contains a photolabile anchoring linkage between a cross-linked polystyrene support and the soluble PEG support, and an acid-labile anchoring bond between the PEG and

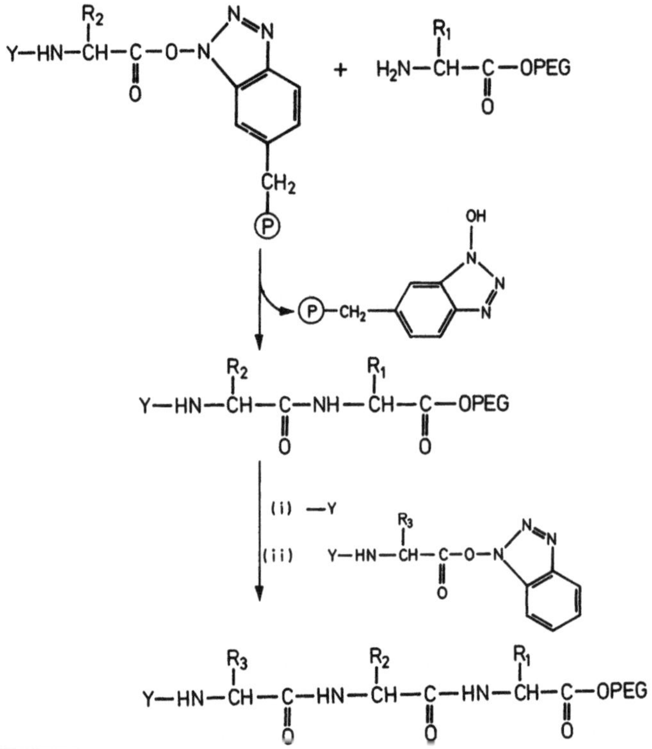

*Scheme 10.* Combination of the Polymer Reagent and Liquid Phase Methods.

the peptide. Such an assembly provides the good physicomechanical characteristics of the cross-linked support and the flexible conformational features of the polyoxyethylene chain, both of which greatly facilitate the synthesis of peptides on the carrier. After the synthesis, the PEG-bound peptide (the preparation of which enables conformational studies of peptides and protein sequences [180] and model studies on enzyme modification [219]), or the free peptide, as desired, can be split of from the polymeric support.

## 5.5 Solubilizing Effect of Polyethyleneglycol Supports and Conformational Analysis of Peptides

In addition to offering a convenient support for the synthesis of a number of model peptides and biologically active protein sequences, the polyethyleneglycol supports also facilitate systematic experimental investigations of the conformations of various peptides and protein sequences. Such investigations have been limited mainly because of the tedious and time-consuming synthetic procedures and the insolubility of most of the sequences in suitable solvents [220]. The problem of the low solubility of peptides is particularly critical because almost all the spectroscopic techniques employed for the conformational studies of peptides demand good solubility in organic and aqueous solvents.

The investigations of peptides and proteins *in aqueous solutions* are of immense relevance as water plays a very important role in the process in which proteins acquire their native structure and in their interaction with other molecules [221,222]. Experimental investigations of the structure of model peptides and protein fragments in aqueous solution are necessary to understand the nature of these interactions. In order to carry out such studies the solubility limitations must be overcome without altering the structure or the intrinsic properties of the peptide sequences.

Attempts have been made to enhance the *solubility properties* of the peptides by copolymerizing them with a hydrophilic peptide sequence [223–226] or by using solubilizing protecting groups [227]. Gratzer and Doty introduced the former technique in order to study sequences which would otherwise be insoluble [223]. They illustrated that a block copolymer, copoly (DL-glutamic acid)$_{175}$-(L-alanine)$_{325}$-(DL-Glutamic acid)$_{175}$, yielded a 100% helical alanine block at pH-values where the glutamates were ionized and conferred solubility. Ingwall et al. [226] investigated the helix-to-coil transition that occurs when poly (L-alanine), incorporated between two blocks of poly(DL-lysine), is subjected to changes in temperature and solvent. Here again the terminal peptide blocks are ionized in the nonhelical form and do not contribute to the optical rotation of the polymers. The method appears to be limited for cases where the specific interactions between the blocks are negligible. Goodman et al. [227] synthesized alanine-containing oligopeptides with 2-methoxy-[2-ethoxy(2-ethoxy)]-acetyl (MEEA) and morpholinamide (Mo) as solubility-enhancing N- and C-terminal protecting groups. However, the solubilizing power of such groups proved not to be strong enough for a general application to many of the difficultly soluble peptide sequences.

A more facile and general approach to overcome the solubility problem is to make use of the *liquid phase synthesis*, since the solubility of the peptide is greatly enhanced by the attachment to the macromolecular C-terminal protecting group, PEG. For the conformational analysis of the bound peptides, the C-terminal polyoxyethylene group must allow the application of all commonly used methods such as CD, NMR, IR, or Raman spectroscopy; at the same time it should not influence the conformational behaviour of the peptide for a direct investigation of the *conformational* properties. PEG shows no Cotton effect and is optically transparent up to the far-UV region; consequently, the measurement of the CD in each step of the synthesis has become a routine procedure in the liquid phase method for following any conformational change of the growing peptide.

The solubility of the PEG-peptides in solvents suitable for CD investigation, like trifluoroethanol, trifluoroacetic acid, or water, is very strong. The infrared investigations can be directly applied to the PEG-bound peptide in the solid state and in solution; the characteristic absorptions of PEG do not interfere with the spectral regions of interest for peptides. The $^{1}$H- and $^{13}$C-NMR of PEG each contain a singlet at $\delta \sim 3.6$ and $\delta \sim 71.5$ ppm; these do not interfere with the signals of the peptide which are sensitive toward conformational changes [210,228,229]. CD spectroscopic analysis of a number of homooligopeptides bound to PEG revealed that the polymeric C-terminal end group does not perturb the peptide conformation from that of the analogous peptides without PEG [228]. This is explained by the conformational properties of PEG; the statistical coil structure with low segmental density excludes any specific interaction between the polyoxyethylene chain and the bound peptide [180].

Thus, the stepwise synthesis on PEG permitted the *systematic analysis* by CD of the strong dependence of the preferred conformations of a number of modelpeptides and biologically active protein sequences on the chain length, solvent polarity, N-protection, side chains, pH, concentration, temperature and ionic interactions [230-234]. Most notably, the attachment to PEG enabled for the first time the CD spectral measurements of the β-forming hydrophobic oligomers of Ala, Val, and Ile upto medium chain lengths [232]. In the case of the PEG-bound oligoalanines, a transition from statistical coil to β-conformation occured at a chain length of 7–8 residues and $(Ala)_7$ showed a complex CD pattern attributable to a mixture of unordered and β-forms. The N-protected PEG-bound $(Ala)_8$ is the first fully protected monodispersed alanine oligopeptide which exhibits a significant amount of ordered secondary structure in water. CD investigations of PEG-bound ACTH sequences revealed a strong tendency for ordered structures at the amino terminus of the peptide [235].

A significant *decrease in the helical structure* was observed when the helix-promoting sequence Pro-Ala-Ala was incorporated at the amino terminus. Transitions from random-coil conformation to partially helical structures, at chain lengths of about 8–10 residues, were observed during the stepwise synthesis of myoglobin-66-73 [168], substance P [168], and model peptides for alamethicin [236]. The strong helicogenic character of the rarely occuring Aib residue became apparent in the study of the alamethicin models; thus, despite the lower intrinsic chirality, the model peptide $(Ala\text{-}Aib)_5$ exhibited ellipticities in the CD spectra similar to those of $(Ala)_{10}$.

The liquid phase synthesis on PEG has also been used for the conformational analysis of *collagen-like sequences* by CD studies [237]. The attachment to PEG has also permitted the CD spectral delineation of the specific interactions between the polypeptide chains and sidechain groups [238,239]. Thus, Anzinger et al. observed that onset of local ordered structures in the mesogenic side chains of polylysine blocks attached to PEG leads to significant, specific alteration in the backbone conformations of the peptide chain [239].

The non-interference of the IR absorption bonds of PEG with the amide A, I, II and V absorptions of the bound peptides and PEG's solubilizing effect allow the *IR conformational analysis* using the liquid phase method. Specifically, the O—H stretching vibration of the free alcoholic groups absorbs strongly only at frequencies higher than 3450 $cm^{-1}$, clearly outside the range where the urethane and amide N—H bonds are observed; the strong C—O—C absorption of polyethyleneglycol is seen at about 1100 $cm^{-1}$. The solubilizing effect of PEG allows the IR investigations on the bound peptides in a wider range of solvents than it is possible with low molecular weight C-terminal esters. IR studies of PEG-bound oligomers of Ala and Val in the solid state indicated that the ordered secondary structure which is developed is the β-structure, characterized by a medium-intensity band at 3280–3260 $cm^{-1}$, a strong band at 1635–1624 $cm^{-1}$, and a weak but distinct band at 720–705 $cm^{-1}$ [240]. This conformation, which is partially formed at the $n = 4$ level, is fully developed at $n = 5$.

The IR study of these peptides *in $CDCl_3$ solution* at high dilution showed that a much higher content of intramolecularly hydrogen-bonded forms exist in $(Ala)_4$ than in $(Val)_4$ peptides. At higher concentrations, chain length, solvent, and side chain effects are all operative in determining the extent of peptide association in these series. IR studies on PEG-bound oligomethionines indicated an essentially unordered conformation upto n = 3 and a partly developed antiparallel β-structure when n = 4–6 [241]. The oligomethionines with n = 7–12 adopt essentially a β-structure and when n = 13–15, the peptides have a significant percentage of α-helical conformation. In the case of PEG-bound oligoglycines, for the observation of a well-developed antiparallel β-structure, a critical chain length of 7 residues was found to be necessary from the IR studies of the PEG-bound analogs [242,243].

*NMR techniques* for the conformational analysis of peptides have been greatly facilitated by the attachment to the PEG support. Owing to the strong solubilizing effect of PEG in the usual NMR solvents, the high concentrations necessary for the measurement of the spectra can be achieved readily. As a result, $^{1}$H-NMR and $^{13}$C-NMR techniques have been applied for the systematic conformational analysis of peptides. Goodman and Saltman investigated the conformational properties of L-glutamate oligopeptides by their stepwise synthesis on a polyethyleneglycol support and delineated the effects of N-terminal end groups and solvents on the conformations of the peptides by high resolution $^{1}$H-NMR [229]. Ribeiro et al. studied the conformational behaviour of a series of oligomethionines by the high resolution 360 MHz $^{1}$H-NMR measurements of the PEG-bound analogs under a variety of conditions [210]. They adopted a combination of high field and selective saturation or Redfield pulse methods [211] for the resolution of individual backbone NH and α-CH resonances of dilute peptides in the presence of the strong resonances from the polyethyleneglycol and/or protonated solvents.

The NMR spectra of these PEG-bound oligomers in $CDCl_3$ are similar to those of the low molecular weight peptide esters, further evidencing other observations that PEG has little effect on the peptide structure. The α-CH region of the peptides is overlapped by signals from the terminal oxyethylene group of PEG, but the peptide side chain and low field backbone NH resonances are well observed. In TFE, the N-protected PEG-bound heptamethionine and octamethionine adopt the right-handed helical structure, and two strong intramolecular hydrogen bonds stabilize these helices; in water, the N-deblocked derivatives form extensive β-sheet structure for n = 5–7, as suggested by the resolved coupling constants $^{3}J_{NH-CH}$ of about 7 Hz for these peptides [210]. The *high-resolution NMR technique* has recently been applied to the conformational analysis of PEG-bound elastin sequences [244].

In the $^{13}$C-NMR spectra, PEG gives a singlet signal at $\delta \sim 71.5$ ppm due to the inner chain carbons [228]. In the conformational studies of PEG-bound sequences of alamethicin, this signal has been eliminated from the spectrum by making use of a double resonance experiment by selective saturation of the methylene carbons [209]. Distinct differences between random-coil and α-helical conformations were observed from these high-resolution $^{13}$C-NMR techniques.

The foregoing typical results clearly illustrate another very important application of the polyethyleneglycol support method for the synthesis of peptides and protein sequences. The unique suitability of this linear, soluble macromolecular support with optimum hydrophilicity-hydrophobicity balance for the conformational analysis of the bound peptides originates from the peculiar conformational properties of the polymer chain.

## 5.6 Conformational Aspects in Peptide Synthesis

The physicochemical properties like the solubility and the rate of coupling play an extremely important role in the stepwise and segment condensation approaches for the synthesis of peptides. In a growing peptide chain, these properties are drastically influenced by conformational transitions. The primary sequence, nature of the involved protecting groups and the physical environment decide the conformational preferences of the concerned peptides. A thorough knowledge of the interdependence of these factors and their influence on the physicochemical properties of the peptides permits an effective planning of the synthetic strategies for the desired peptide and protein sequences. The liquid phase synthesis of model oligopeptides on the polyethyleneglycol support and subsequent investigation of the conformational preferences with respect to the sequence, protecting groups, chain length, and solvation on secondary structure help to delineate the above-mentioned interdependence. The resulting correlation between conformation, solubility and coupling kinetics of the growing peptide chain can be made use of for the design of a specific experimental strategy for a successful synthesis of peptides and protein sequences.

### 5.6.1 Chain Elongation and Physicochemical Properties

The solubility of the peptides and reactivity of the terminal amino group in the coupling reaction in the LPM are significantly reduced at stages where the oligomers exhibit β-structures [231]. Thus, in the PEG-bound homooligomers of Val and Ile, the tendency to aggregate at very short chain lengths is evident from the drastic decrease in solubility with the growing peptide chain. For example, when dissolved in methylene-chloride, the tetrapeptide derivatives yield highly viscous solutions at concentrations of 10% (w/v), and only at very low concentrations ($<1\%$) the viscosity decreases to normal values. Obviously, a network-like structure is formed by intermolecular interactions between the peptide chains. By the addition of DMF or DMSO, these aggregates are disrupted and the viscosity decreases.

With increasing chain length, increasing amounts of DMF or DMSO had to be used to obtain homogeneous solutions. The PEG-bound $(Ile)_7$ and $(Ile)_8$ were only sparingly soluble even in mixtures of DMF and DMSO, and resulted in gel-like suspensions in $CH_2Cl_2$ even at concentrations as low as 1%. Whereas the low solubility of these oligopeptides represent the most severe obstacle in the stepwise elongation, quantitative yields are also difficult to achieve due to the reduced reactivity of the terminal amino groups. CD-studies indicated that this change in the physicochemical properties is paralleled by a conformational transition from random-coil (r.c.) to β-structure, and the unusual insolubility of $(Ile)_8$ is explained by the possibility of the formation of an intramolecular β-structure (cross β-structure).

Oligopeptides with tendencies to adopt α-helical or unordered structures showed no pronounced change in solubility or coupling kinetics during chain elongation in the LPM as exemplified in the case of the oligomethionines [181]. For example, the oligomethionine, L-$(Met)_{15}$-NH-PEG exhibits excellent solubility in a number of solvents (Table 5).

In the case of Glu(OBzl), the solubility of the oligomers in $CH_2Cl_2$ decreased considerably with the growing peptide chain at the beginning and showed minimum

**Table 5.** Conformation and Solubility of Peptide-PEG esters

| PEG[a]-bound Peptide | Conformation in trifluoroethanol | Solubility[b] in | | |
|---|---|---|---|---|
| | | TFE | $CH_2Cl_2$ | DMF |
| Boc-(L-Met)$_n$; n > 10 | α | ++ | + | + |
| (L-Met)$_5$-Pro-(L-Met)$_5$ | r.c. | + | o | + |
| [L-Glu(OBzl)]$_8$ | β | o | — | — |
| [L-Glu(OBzl)]$_{14}$ | α | ++ | + | ++ |
| [L-Glu(OBzl)]$_7$-Pro-[L-Glu(OBzl)]$_7$ | r.c.(α) | ++ | + | ++ |
| [L-Glu(OBzl)]$_2$-Pro-[L-Glu(OBzl)]$_{12}$ | α | ++ | + | ++ |
| [L-Glu(OBzl)]$_{20}$-Gly | α | ++ | + | ++ |
| (L-Val)$_8$-Gly | β | o | —— | — |
| (L-Ile)$_n$; n > 6 | β | — | —— | —— |
| (L-Ile)$_4$-Gly-(L-Ile)$_4$ | r.c.(β) | + | o | + |
| (L-Ile)$_4$-Pro-(L-Ile)$_4$ | r.c.(β) | + | o | + |
| (L-Ile)$_n$-X; (X = Gly, Pro); n > 6 | β | — | —— | —— |
| (L-Ala)$_n$; n > 7 | β | — | —— | — |
| (L-Ala)$_5$-Pro-(L-Ala)$_4$ | r.c. | + | o | + |

*a* Molecular weight 10,000; *b* ++, very good; +, good; —, Low; ——, very low; o, medium

solubility in the range n = 6–8. With further elongation of the peptide chain a continuous increase in the solubility was observed, and for n > 12, the oligopeptide exhibited the same solubility as that of the lowest oligomers (n < 5). This change in solubility with chain length is paralleled by conformational transitions of the type, random-coil (n ≦ 5) → β (5 < n ≦ 8) → α-helix [230], as depicted in Fig. 2. The conformational features outlined above were also reflected in the kinetics of the coupling processes. For example, a considerable decrease in the reaction rates was observed in the range 3 < n < 9, indicating a reduced reactivity of the terminal amino group of the oligomers in the β-conformations. On proceeding to increased chain lengths, transition to α-helical conformations occured and a significant increase in the reaction rate was registered.

A better understanding of the relation between physicochemical properties and conformation of oligopeptides has been possible by the application of the so-called host-guest technique in the liquid phase method of peptide synthesis. This technique was originally introduced to obtain detailed information regarding the thermo-

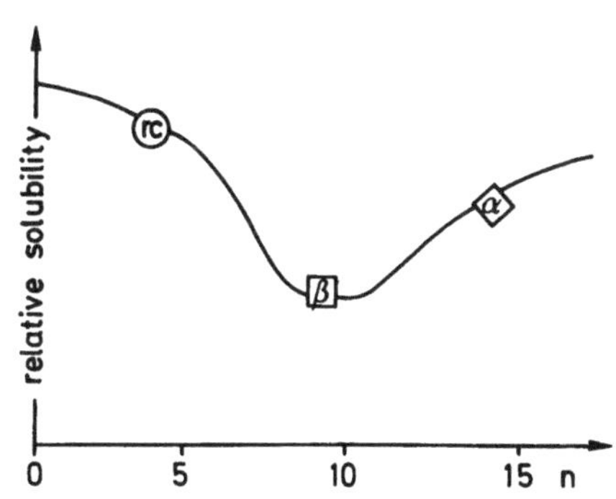


**Fig. 2.** Solubility and conformation of oligopeptides

dynamic and conformational parameters of single guest amino acids in various positions of a host peptide chain [245,246]. Considering conformational energies, the insertion of Gly or Pro residue should result in a substantial disruption of ordered conformations. This is evident from the investigation of the conformation and physicochemical properties of a number of co-oligopeptides ("host-guest peptides") bound to PEG. CD investigations show that the α-helix content in TFE of PEG-bound Boc-$(Met)_5$-Pro-$(Met)_5$ (28%) approaches that of Boc-$(Met)_7$ (30%) and it is much lower than that of Boc-$(Met)_{11}$ (80%) [247]. Similarly, the CD spectrum of PEG-bound Boc-$(Ala)_{10}$ is indicative of a pure β-structure, whereas Boc-$(Ala)_5$-Pro-$(Ala)_4$-PEG gives a spectrum characteristic of mostly unordered conformations [248] (Fig. 3).

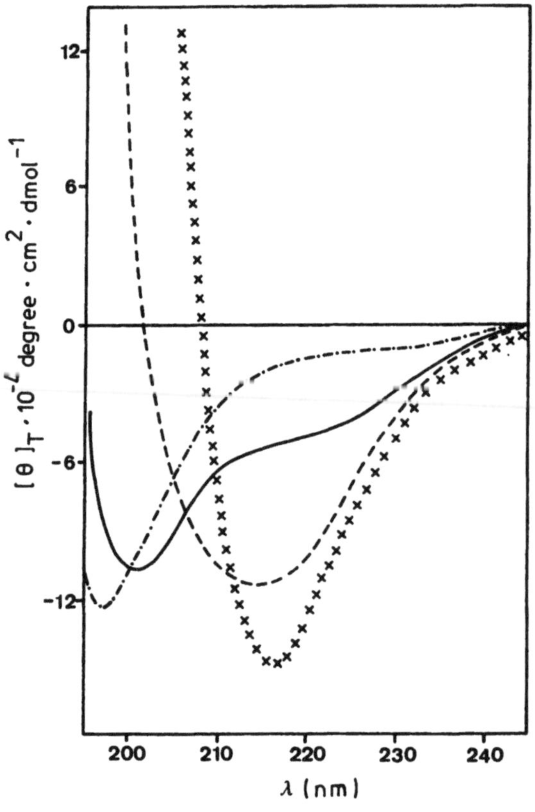


**Fig. 3.** CD-Spectra of PEG-bound Boc-$(Ala)_{10}$ in water (× × × ×) and in TFE (– – – –); Boc-$(Ala)_5$-Pro-$(Ala)_4$ in water (–·–·–·) and in TFE (———)

The drastic effect of the Pro residue on the ordered conformations is also demonstrated for the case of PEG-bound Boc-L-Glu(OBzl) co-oligopeptides in the solid state and in solution [248,249].

The observed effect of Pro and Gly on the conformations of the host oligopeptides are paralleled by drastic changes in the solubility and coupling rates of the peptides [181]. Thus, the solubility of $(Ile)_4$-Gly-$(Ile)_4$ and $(Ile)_4$-Pro-$(Ile)_4$ differ drastically from the corresponding homooligomer. The Ile-Pro "host-guest peptide" is soluble in $CH_2Cl_2$, DMF or AcOH, whereas the homooligomer is practically insoluble in these solvents. The corresponding glycine-substituted co-oligomer shows somewhat lower solubility compared to the proline-substituted one, but here also the difference in the physicochemical properties compared to the homooligopeptide is still striking. The coupling rates also remain practically constant throughout the stages of chain elongation in the case of the host-guest peptide. For example, the reaction condition for obtaining quantitative yields for the steps $(Ile)_5 \dashrightarrow (Ile)_8$ did not differ significantly from those in the steps $(Ile)_1 \dashrightarrow (Ile)_4$ in the case of $(Ile)_4$-Gly-$(Ile)_4$ or $(Ile)_4$-Pro-$(Ile)_4$. This is in sharp contrast to that observed in the case of the homooligopeptide.

### 5.6.2 Implications in Synthetic Strategies

The foregoing results on the interrelation between the conformation and physicochemical properties of the peptides are of much significance in chosing an effective synthetic strategy for a particular sequence. The above results suggest that the elongation of the peptide chain must not necessarily be paralleled by a decrease in solubility. The observation that medium-sized peptides show minimum solubility and reactivity is related to the tendency of these short segments to form β-like structures and this behaviour appears quite general. However, this critical stage in peptide synthesis can be overcome either by

1. further elongation of the peptide chain (i.e. in cases where structures other than β-conformation become more favourable at longer chain lengths), most notably by segment condensation, or by
2. insertion of a residue, which prevents the formation of β-structure, in the middle of medium-sized peptide chain.

These findings can be of considerable impact for the conventional strategies of peptide synthesis. The segment condensation of medium-sized peptides with low solubility to larger peptide chains may result in the increased solubility of the reaction product due to conformational transitions. Whenever it is possible, amino acid residues with low tendencies for β-structure formation should be inserted towards the middle of the peptide segments. In this respect, the prediction of secondary structures can be a useful tool in establishing the most efficient path for peptide synthesis. To this end, experimental elucidation of the conformational preferences of side-chain-protected trifunctional amino acids is of much relevance [250].

Thus, considering these conformational features, an effective synthetic strategy for larger peptides can be envisioned as follows: first, determine the nature of the secondary structure for the various possible segments following empirical prediction rules [251,252]. Then, arrange the sequence in such a way as to avoid longer $(n > 5)$ segments of amino acid residues with $P_\beta$-potentials larger than 1. If necessary and when the sequence permits, insert a β-breaking residue in a central position of the concerned segment; proceed to the synthesis of these segments by the stepwise approach and finally establish the total sequence by segment condensation. Such a procedure which gives due consideration to the conformational transitions and the dependent physicochemical properties of the peptides should help to overcome two main impediments in peptide synthesis, namely the low solubility of segments and the reduced reactivity of the terminal amino groups with increasing chain length.

## 6 Conclusions and Outlook

Recent advances made by a number of research groups in the improvement of the polymeric supports, coupling methods, anchoring and protecting groups, and automation have enabled the solid phase peptide synthesis, originally introduced by Merrifield approximately two decades ago, to contribute significantly to peptide and protein synthesis. Chemical problems associated with the repetitive stepwise synthesis on the crosslinked polystyrene support have been partially overcome by several methodological refinements in the functionalization reactions, anchoring groups and reaction protocols. Recent detailed understanding of the solvation, swelling and solubilizing properties of the polymeric supports have led to significantly improved synthetic approaches in the solid phase methodology.

The *segment condensation approach* is now one of the rapidly developing techniques in the solid phase method. The possibility of the isolation, purification and characterization of the intermediates in this approach is definitely a notable advantage over the stepwise synthesis on the polymeric support.

The *physicochemical incompatibility* of the polystyrene supports with the peptides has been observed to be one of the problems associated with the original Merrifield method. This has been approached successfully by Sheppard and coworkers by the introduction of the polar *polyacrylamide supports*. Poly(*N*-acrylylpyrrolidine) resin is another polar polymeric support developed on similar grounds. The use of these polar resin supports facilitated the solid phase synthesis of a number of peptides and protein sequences in higher yield and purity than with the polystyrene resins. However, the inadequate availability of these polar supports appears to be a limiting factor for a wider utilization at present.

The polymer reagent method of peptide synthesis, which makes use of *insoluble polymeric active esters*, is suited for the synthesis of *short peptides* in higher purity than is possible with the Merrifield method. The investigation of the suitability of the heterogeneous polymeric networks for biopolymer synthesis has initiated much studies directed towards the understanding of the reactivity parameters in polymeric matrix reactions. The role of the degree of crosslinking and topographical nature in deciding the reactivity of functional groups attached to the heterogeneous matrix has been investigated. The non-equivalence of functional groups was observed to be one of the reasons for the incomplete reactions in solid polymeric matrices.

From a reactivity point of view, the use of the soluble polymeric techniques is a significant variation to the heterogeneous solid phase methods. Among the soluble polymers so far investigated, *polyethyleneglycol* is the best suited one for the *stepwise synthesis* of biopolymers, particularly of peptides. In the Liquid Phase Method (LPM), PEG is used as a solubilizing macromolecular C-terminal protecting group. The method permits rapid synthesis of peptides in solution and is a definite simplification of the experimental procedures involved in the classical peptide synthesis. The reactions carried out in *homogeneous solutions* go to completion and the soluble polymeric protecting group has no unfavourable influence on the attached peptides. However, the solubility problems at longer chain lengths in the case of certain sequences set limits on the application of the method. The synthesis on the PEG support is possible as long as the physicochemical properties of the polymer-bound peptide are dominated by the polymer. In the case of *helix-forming and randomly coiling peptides*, the solubility problem does not arise in the LPM, whereas in the case of β-forming sequences, a critical drop in solubility occurs at very short chain lengths. This problem of aggregation resulting in the reduced solvation of the peptide chains, which is characteristic of all the solution methods, appears less severe in the solid phase synthesis on very lightly crosslinked polystyrene matrices.

Another promising application of the synthesis on PEG supports is in the facilitation of the *conformational analysis* of peptides and protein sequences. Due to the optimum hydrophilic-hydrophobic balance, the polymer chain exerts a strong solubilizing effect on the attached peptides in a wide variety of solvents and this helps the application of the various physicochemical methods for conformational investigations. This method has permitted the delineation of the impact of conformation on the synthetic strategies for peptide sequences. The solubility of the PEG-bound peptides in aqueous systems can be made use of for the investigation of protein sequences in *aqueous solutions*. Such studies are of immense relevance as water plays a very important role in the process in which proteins acquire their native structure and in their interaction with other small and large molecules. The investigation of PEG-peptides as models for biologically active compounds can be envisioned as a fruitful area for further research in this regard.

An analysis of the characteristics of the different insoluble and soluble supports developed for peptide synthesis gives the clue to criteria for a *most suitable macromolecular support*. The ease of handling of the polymeric carrier and the high reactivity of the functional groups attached to them are the deciding factors in this case. However, an ideal combination of these factors remains difficult to attain. In this connection, attempts to combine the advantages of the insoluble and soluble polymeric supports could lead to the design of improved, more versatile macromolecular supports for peptide synthesis. Thus, a combination of strategies rather than a single strategy appears best suited for the synthesis of this class of complex biological macromolecules [253]. In this respect also, peptide synthesis continues to remain an "undiminished" [254] and "formidable" [255] challenge. The biological principles of the stepwise condensation of amino acids occuring in the gelatinous phase must also be considered for optimization of the existing chemical methods on macromolecular supports, which differ drastically from the far more rapid reactions in protein biosynthesis.

## 7 Acknowledgements

We thank the Deutsche Forschungsgemeinschaft for the financial support of parts of our work described in this article through the Sonderforschungsbereich 41, "Chemie und Physik der Makromoleküle".

## 8 References

1. Merrifield, R. B.: J. Am. Chem. Soc. *85*, 2149 (1963)
2. Gutte, B., Merrifield, R. B.: J. Am. Chem. Soc. *91*, 501 (1969)
3. Li, C. H., Yamashiro, D.: J. Am. Chem. Soc. *92*, 7608 (1970)

4. For reviews see: Kössel, H., Seliger, H.: Progr. Chem. Org. Nat. Prod. *32*, 430 (1975); Amaranath, V., Broom, A. D.: Chem. Rev. *77*, 209 (1977)
5. Frechet, J. M. J., Schuerch, C.: J. Am. Chem. Soc. *93*, 492 (1971); Frechet, J. M. J., Schuerch, C.: Carbohyd. Res. *22*, 399 (1972); Hanessian et al.: Carbohyd. Res. *38*, 15 (1974); Seymour, E., Frechet, J. M. J.: Tetrahedron Lett. 1976, 1149; Zehavi, U., Amit, B., Patchornik, A.: J. Org. Chem. *37*, 2281 (1972); Zehavi, U., Patchornik, A.: J. Am. Chem. Soc. *95*, 5673 (1973)
6. Reviews: Leznoff, C. C.: Acc. Chem. Res. *11*, 327 (1978); Frechet, J. M. J.: Tetrahedron *37*, 663 (1981); Kraus, M. A., Patchornik, A.: J. Polym. Sci., Macromol. Rev. *15*, 55 (1980); Akelah, A.: Synthesis, *1981*, 413; Overberger, C. G., Sannes, K. N.: Angew. Chem. *86*, 139 (1974); Angew. Chem. Int. Ed. Engl. *13*, 99 (1974); Mathur, N. K., Williams, R. E.: J. Macromol. Sci. Rev. Macromol. Chem. *C15*, 117 (1976)
7. Tsuchida, E., Nishide, H.: Adv. Polym. Sci. *24*, 1 (1977)
8. Manecke, G., Storck, W.: Angew. Chem. *90*, 691 (1978); Angew. Chem. Int. Ed. Engl. *17*, 657 (1978); Goldstein, L., Manecke, G.: in: Applied Biochem. and Bioeng., (Eds. Wingard, L. B., Katchalski-Katzir, E.; Goldstein, L.) Academic Press, New York 1976, p. 23; Royer, G. P.: Catal, Rev. *22*, 29 (1980)
9. Okawara, M. et al.: J. Macromol. Sci. Chem. *A13*, 441 (1979); Sherrington, D. C.: Brit. Polym. J. *12*, 70 (1980)
10. Review: Rabek, Jr., J.: Tetrahedron *35*, 723 (1979)
11. Erickson, B. W., Merrifield, R. B.: Solid-phase peptide Synthesis, in: The Proteins, (Eds., Neurath, H., Hill, R. L.), Vol. II, Academic Press, New York 1976³ p. 259; Barany, G., Merrifield, R. B.: solid-phase peptide synthesis, in: The Peptides, (Eds. Gross, E., Meienhofer, J.), Vol. 2, Academic Press, New York 1979, p. 3
12. Birr, C.: Aspects of the Merrifield Peptide Synthesis, Springer, Berlin-Heidelberg-New York 1978
13. Heitz, W.: Adv. Polym. Sci. *23*, 1 (1977)
14. Heitz, W.: Angew. Makromol. Chem. *76*, 273 (1979)
15. Dally, W. H.: Makromol. Chem., Suppl. *2*, 3 (1979)
16. Bayer, E. et al.: J. Am. Chem. Soc. *92*, 1735 (1970)
17. Hirt, J., de Leer, E. W., Beyerman, H. C.: In: The Chemistry of Polypeptides, (Ed. Katsoyannis, P. G.), Plenum, New York 1973, p. 363; Hodges, R. S., Merrifield, R. B.: Anal. Biochem. *65*, 241 (1975); Brunfeldt, K.: Acta Physica Chemica, *23*, 499 (1977)
18. Birr, C.: Liebigs Ann. Chem. *763*, 162 (1972)
19. Birr, C.: Liebigs Ann. Chem. *1973*, 1652
20. Birr, C.: in: Peptides 1974, (Ed. Wolman, Y.), Wiley, New York 1975, p. 117
21. Gut, V., Rudinger, J.: In: Peptides 1968, (Ed. Bricas, E.), North Holland Publ., Amsterdam 1968, p. 185
22. Gut, V.: Collect. Czech. Chem. Commun. *40*, 129 (1975)
23. Wang, S. S., Merrifield, R. B.: Int. J. Protein Res. *1*, 235 (1969)
24. Weygand, F., Obermeier, R.: Z. Naturforsch. *B23*, 1390 (1968)
25. Geising, W., Hörnle, S.: in: Peptides 1971, (Ed. Nesvadba, H.), North Holland Publ., Amsterdam 1973, p. 146
26. Niall, H. D., Tregear, G. W., Jacobs, J.: in: Chemistry and Biology of Peptides, (Ed. Meienhofer, J.), Ann Arbor Sci. Publ., Ann Arbor-Michigan 1972, p. 695
27. Udenfried, S. et al.: Science *178*, 871 (1972)
28. Felix, A. M., Jimenez, M. H.: Anal. Biochem. *52*, 377 (1973)
29. Brunfeldt, K., Roepstorff, P., Thomsen, J.: Acta Chem. Scand. *23*, 2906 (1969)
30. Brunfeldt, K., Christensen, T., Villemoes, P.: FEBS Lett. *22*, 238 (1972)
31. Merrifield, R. B., Stewart, J. M.: Nature (London) *207*, 522 (1965)
32. Merrifield, R. B., Stewart, J. M., Jernberg, N.: Anal. Chem. *38*, 1905 (1966)
33. Brunfeldt, K.: in: Peptides 1972, (Eds. Hanson, H., Jakubke, H. D.), North Holland Publ., Amsterdam 1973, p. 141
34. Gisin, B. F.: Anal. Chim. Acta *58*, 248 (1972)
35. Gisin, B. F., Merrifield, R. B.: J. Am. Chem. Soc. *94*, 3102 (1972)
36. Hodges, R. S., Merrifield, R. B.: Anal. Biochem. *65*, 241 (1975)
37. Villemoes, P., Christensen, T., Brunfeldt, K.: Hoppe-Seyler's Z. Physiol. Chem. *357*, 713 (1976)

38. Villemoes, P., Christensen, T., Brunfeldt, K.: Acta Chem. Scand. *B32*, 703 (1978)
39. Edelstein, M. S. et al.: Abstr. Am. Pept. Symp. 7th, 91 (1981)
40. Merrifield, R. B.: Beckman Rep. *1972*, 3; Chem. Abstr. *77*, 102155 m (1972)
41. Westfall, F. C., Robinson, A. B.: J. Org. Chem. *35*, 2842 (1970)
42. Hagenmaier, H.: Tetrahedron Lett. *1970*, 283
43. Chou, F. C. H. et al.: J. Am. Chem. Soc. *93*, 267 (1971)
44. Sheppard, R. C.: in: Peptides 1971, (Ed. Nesvadba, H.), p. 111, North Holland Publ., Amsterdam 1973, p. 111
45. Frankhauser, P., Brenner, M.: in: The Chemistry of Polypeptides, (Ed. Katsoyannis, P. G.), Plenum, New York 1973, p. 389
46. Merrifield, R. B.: Adv. Enzymol. *32*, 221 (1969)
47. Meienhofer, J.: in: Hormonal Proteins and Peptides, (Ed. Li, C. H.), Academic Press, New York 1973, p. 45
48. Anfinsen, C. B.: in: Chemistry of Natural Products, IUPAC-Butterworths, London 1968, p. 461
49. Chapman, T. M., Kleich, D. G.: J. Chem. Soc., Chem. Commun. *1973*, 193
50. Atherton, E., Sheppard, R. C.: in: Peptides 1974, (Ed. Wolman, Y.), Wiley, New York 1975, p. 123
51. Arshady, R. et al.: J. Chem. Soc., Chem. Commun. *1979*, 423
52. Atherton, E. et al.: J. Am. Chem. Soc. *97*, 6584 (1975)
53. Atherton, E. et al.: in: Peptides 1976, (Ed. Loffet, A.), Univ. of Brussels, Brussels 1976, p. 291
54. Atherton, E., Sheppard, R. C.: in: Peptides — Proc. Am. Pept. Symp. 5th., (Eds. Goodman, M., Meienhofer, J.), Wiley, New York 1977, p. 503
55. Atherton, E. et al.: J. Chem. Soc. Chem. Commun. *1978*, 539
56. Inman, J. K., Dintzis, H. M.: Biochem. *8*, 4074 (1969)
57. Arshady, R. et al.: J. Chem. Soc., Chem. Commun. *1979*, 423
58. Atherton, E. et al.: Bioorg. Chem. *8*, 351 (1979)
59. Arshady, R. et al.: J. Chem. Soc., Perkin I *1981*, 529
60. Atherton, E., Logan, C. J., Sheppard, R. C.: J. Chem. Soc., Perkin I *1981*, 538
61. Gait, M. J., Sheppard, R. C.: Nucleic Acids Res. *4*, 1135 (1977)
62. Atherton, E., Bridgen, J., Sheppard, R. C.: FEBS Lett. *64*, 173 (1976)
63. Atherton, E., Brown, E., Sheppard, R. C.: J. Chem. Soc., Chem. Commun. *1981*, 1151
64. Smith, C. W., Stahl, G. L., Walter, R.: Int. J. Pept. Protein Res. *13*, 109 (1979)
65. Stahl, G. L., Walter, R., Smith, C. W.: J. Am. Chem. Soc. *101*, 5383 (1979)
66. Glass, J. D., Walter, R., Schwartz, I. L.: in: Peptides 1972, (Eds. Hanson, H., Jakubke, H. D.), North Holland Publ., Amsterdam 1973, p. 135
67. Glass, J. D. et al.: J. Am. Chem. Soc. *96*, 6476 (1974)
68. Glass, J. D., Schwartz, I. L., Walter, R.: J. Am. Chem. Soc. *94*, 6209 (1972)
69. Smith, C. W., Skala, G., Walter, R.: Int. J. Pept. Protein Res. *16*, 365 (1980)
70. Weygand, F., Ragnarsson, U.: Z. Naturforsch. *B21*, 1141 (1966)
71. Omenn, G. S., Anfinsen, C. B.: J. Am. Chem. Soc. *90*, 6571 (1968); Yajima, H., Kawatani, H.: Chem. Pharm. Bull. *19*, 1905 (1971); Matsueda, R. et al.: Bull. Chem. Soc. Jap. *46*, 3240 (1973); Karlsson, S. M., Ragnarsson, U.: Acta Chem. Scand. *B28*, 376 (1974); Urry, D. W., Cunningham, W. D., Ohnishi, T.: Biochem. *13*, 609 (1974); Weber, U., Andre, M.: in: Peptides 1974, (Ed. Wolman, Y.), Wiley, New York 1975, p. 153; Ragnarsson, U., Karlsson, S. M., Hamburg, U.: Int. J. Pept. Protein Res. *7*, 307 (1975); Larsson, L. E., Melin, P., Ragnarsson, U.: Int. J. Pept. Protein Res. *8*, 39 (1976)
72. Yajima, H., Kawatani, H., Watanabe, H.: Chem. Pharm. Bull. *18*, 1333 (1970)
73. Wang, S. S., Merrifield, R. B.: Int. J. Pept. Protein Res. *4*, 309 (1972)
74. Wang, S. S.: J. Am. Chem. Soc. *95*, 1328 (1973)
75. Birr, C., Müller, M. W., Buku, A.: in: Peptides — Proc. Am. Pept. Symp. 5th., (Eds. Goodman, M., Meinehofer, J.), Wiley, New York 1977, p. 510
76. Birr, C.: Abstr. Am. Pept. Symp. 7th, p. 87 (1981)
77. Birr, C.: in: Ref. 12, p. 57; Sato, K. et al.: Bull. Chem. Soc. Jap. *50*, 1999 (1977)
78. Yajima, H., Kiso, Y.: Chem. Pharm. Bull. *22*, 1087 (1974)
79. Yajima, H. et al.: J. Chem. Soc. Chem. Commun. *1974*, 106

80. Maruyama et al.: Bull. Chem. Soc. Jap. *49*, 2259 (1976); Kawatani, H., Tamura, F., Yamija, J.: Chem. Pharm. Bull. *22*, 1879 (1974)
81. Weber, U., Andre, M.: Z. Physiol. Chem. *356*, 701 (1975)
82. Protein Synthesis Group — Shanghai Inst. of Biochem.: Scientia Sinica *18*, 745 (1975)
83. Bhatnagar, R. S., Rapaka, P. S.: Biopolymers *14*, 597 (1975)
84. Rapaka, R. S., Bhatnagar, R. S.: Int. J. Pept. Protein Res. *7*, 475 (1975)
85. Niu, C. I. et al.: Abstr. Am. Pept. Symp. 7th., p. 87 (1981)
86. Atherton, E., Sheppard, R. C.: Abstr. Am. Pept. Symp. 7th., p. 88 (1981)
87. Tam, J. P., Tjoeng, F. S., Merrifield, R. B.: Tetrahedron Lett. *1979*, 4935
88. Tam, J. P., Tjoeng, F. S., Merrifield, R. B.: in: Peptides — Proc. Am. Pept. Symp. 6th., (Eds. Gross, E., Meienhofer, J., Vigna, R.), Pierce Chemical Co., Illinois 1979, p. 341
89. Tam, J. P., Tjoeng, F. S., Merrifield, R. B.: J. Am. Chem. Soc. *102*, 6117 (1980)
90. Tam, J. P., Dimarchi, R. D., Merrifield, R. B.: Int. J. Pept. Protein Res. *16*, 412 (1980); Tam, J. P.: Abstr. Am. Pept. Symp. 7th. p. 88 (1981)
91. Merrifield, R. B. et al.: in: Peptides — Proc. Am. Pept. Symp. 5th., (Eds. Goodman, M., Meienhofer, J.), Wiley, New York 1977, p. 488
92. Barany, G., Merrifield, R. B.: J. Am. Chem. Soc. *99*, 7363 (1977)
93. Fridkin, M., Patchornik, A., Katchalski, E.: Isr. J. Chem. *3*, 69 (1965)
94. Fridkin, M., Patchornik, A., Katchalski, E.: J. Am. Chem. Soc. *88*, 3164 (1966)
95. Wieland, T., Birr, C.: Angew. Chem. *78*, 303 (1966); Angew. Chem., Int. Ed. Engl. *5*, 310 (1966)
96. Wieland, T., Birr, C.: Chimia *21*, 581 (1967)
97. Sklyarov, L. Y., Shashkova, I. V.: Zh. Obshch. Khim. *39*, 2788 (1969)
98. For a review see: Fridkin, M.: Polymeric reagents in peptide synthesis, in: The Peptides (Eds. Gross, E., Meienhofer, J.), Vol. 2, Academic Press, New York 1979, p. 333
99. Kalir, R., Fridkin, M., Patchornik, A.: Eur. J. Biochem. *42*, 151 (1974)
100. Tsiryapkin, V. A. et al.: Dokl. Akad. Nauk SSR *223*, 1156 (1975)
101. Andreev, S. M. et al.: Synthesis 303 (1977)
102. Teramato, T., Narita, M., Okawara, M.: J. Polym. Sci., Polym. Chem. Ed. *15*, 1369 (1977)
103. Lauren, D. R., Williams, R. E.: Tetrahedron Lett. 2665 (1972)
104. Wieland, T., Birr, C.: in: Peptides 1966 (Eds. Beyerman, H. C. et al.), North Holland Publ., Amsterdam 1967, p. 103
105. Stern et al.: J. Solid-Phase Biochem. *2*, 131 (1977)
106. Fridkin, M., Patchornik, A., Katchalski, E.: J. Am. Chem. Soc. *90*, 2953 (1968)
107. Kalir, R. et al.: Eur. J. Biochem. *59*, 55 (1975)
108. Fridkin, M. et al.: J. Solid-Phase Biochem. *2*, 175 (1977)
109. Fridkin, M., Patchornik, A., Katchalski, E.: J. Am. Chem. Soc. *87*, 4646 (1965)
110. Patchornik, A., Fridkin, M., Katchalski, E.: in: Peptides 1966 (Eds. Beyerman, H. C. et al.), North Holland Publ., Amsterdam 1967, p. 91
111. Fridkin, M., Patchornik, A., Katchalski, E.: Biochem. *11*, 466 (1972)
112. Morawetz, H.: Pure Appl. Chem. *51*, 2307 (1979)
113. Pan, S. S., Morawetz, H.: cited in ref. 112
114. Losse, A.: Tetrahedron Lett. *1971*, 4989
115. Losse, A.: Tetrahedron *29*, 1203 (1973)
116. Frank, H., Hagenmaier, H.: Tetrahedron *30*, 2523 (1974)
117. Rudinger, J., Gut, V.: in: Peptides 1967, (Eds. Beyerman, H. C. et al.), North Holland Publ., Amsterdam 1967, p. 89
118. Rudinger, J., Beutzer, P.: in: Peptides 1974, (Ed. Wolman, Y.), Wiley, New York 1975, p. 211
119. Andreatta, R. H., Rink, H.: Helv. Chim. Acta *56*, 1205 (1973)
120. Losse, G., Ulbrich, R.: Tetrahedron *28*, 5823 (1972)
121. Gut, V., Rudinger, J.: Colloq. Int. C.N.R.S. *175*, 185 (1968)
122. Maher, J. J., Furey, M. E., Grennberg, L. J.: Tetrahedron Lett. *1971*, 27
123. Kau, J. I., Morawetz, H.: Polym. Prepr. Polym. Chem. Div. Am. Chem. Soc. *13*, 819 (1972)
124. Morawetz, H.: J. Polym. Sci., Polym. Symp. *62*, 271 (1978)

125. Letsinger, R. L., Kornet, M. J.: J. Am. Chem. Soc. *85*, 3045 (1963)
126. Letsinger, R. L., Jerina, D. M.: J. Polym. Sci. Part A-1, *5*, 1977 (1967)
127. Letsinger, R. L. et al.: J. Am. Chem. Soc. *86*, 5163 (1964)
128. Frechet, J. M., Farral, M. J.: in: Chemistry and Properties of Crosslinked Polymers, (Ed. Labama, S. S.), Academic Press, New York 1977, p. 59
129. Tilak, M. A., Hollinden, C. S.: Org. Prep. Proced. Int. *3*, 183 (1971)
130. Sano, S., Tokunaga, R., Kun, K. A.: Biochim. Biophys. Acta *244*, 201 (1971)
131. Sano, S.: in: Oxidation Reduction Enzymes, (Ed. Akeson, A., Ehrenberg, A.), Pergamon Press, Oxford 1972, p. 35
132. Regen, S. L., Lee, D. P.: J. Am. Chem. Soc. *96*, 294 (1974)
133. Regen, S. L.: J. Am. Chem. Soc. *96*, 5275 (1974)
134. Regen, S. L.: Macromolecules *8*, 689 (1975)
135. Lloyd, W. G., Alfrey, Jr. T.: J. Polym. Sci. *62*, 159 (1962)
136. Lloyd, W. G., Alfrey, Jr., T.: J. Polym. Sci. *62*, 301 (1962)
137. Crowley, J. I., Rapoport, H.: Acc. Chem. Res. *9*, 135 (1976)
138. Bonds, W. D. et al.: J. Am. Chem. Soc. *97*, 2128 (1975)
139. Jayalekshmy, P., Mazur, S.: J. Am. Chem. Soc. *98*, 6710 (1976)
140. Crowley, J. I., Harvey, T. B., Rapoport, H.: J. Macromol. Sci. Chem. *7*, 1118 (1973)
141. Scott, L. I. et al.: J. Am. Chem. Soc. *99*, 625 (1977)
142. Kraus, M. A., Patchornik, A.: J. Am. Chem. Soc. *93*, 7325 (1971)
143. Beyerman, H. C., de Leer, E. W. B., van Vossen, W.: J. Chem. Soc. Chem. Commun. *1972*, 929
144. Rothe, M., Mazanek, J.: Tetrahedron Lett. *1972*, 3795
145. Rothe, M., Mazanek, J.: Angew. Chem. *84*, 290 (1972); Angew. Chem. Int. Ed. Engl. *11*, 293 (1972)
146. Rothe, M., Mazanek, J.: in: Chemistry and Biology of Peptides, (Ed. Meienhofer, J.), Ann Arbor Sci. Publ., Ann Arbor-Michigan 1972, p. 89
147. Rothe, M., Mazanek, J.: Liebigs Ann. Chem. *1974*, 439
148. Rothe, M. et al.: in: Peptides — Proc. Am. Pept. Symp. 5th., (Eds. Goodman, M., Meienhofer, J.), Wiley, New York 1977, p. 506
149. Lunkenheimer, W., Zahn, H.: Liebigs Ann. Chem. *740*, 1 (1970)
150. Lunkenheimer, W., Zahn, H.: Angew. Makromol. Chem. *10*, 69 (1970)
151. Gutte, B., Merrifield, R. B.: J. Biol. Chem. *246*, 1922 (1973)
152. Wieland, T., Birr, C., Flor, F.: Liebigs Ann. Chem. *727*, 130 (1969)
153. Marshall, G. R., Merrifield, R. B.: in: Biochemical Aspects of Reactions on Solid Supports, (Ed. Stark, G. R.) p. 111, Academic Press, New York 1971, p. 111
154. Sarin, V. K., Kent, S. B. H., Merrifield, R. B.: J. Am. Chem. Soc. *102*, 5463 (1980)
155. Krause, S.: J. Polym. Sci. *35*, 558 (1959)
156. Birr, C.: ref. 12, p. 18
157. Geckeler, K., Pillai, V. N. R., Mutter, M.: Adv. Polym. Sci. *39*, 65 (1981)
158. Shemyakin, M. M. et al.: Tetrahedron Lett. *1965*, 2323
159. Ovchinnikov, Y. A., Kiryushkin, A. A., Kozhevnikova, I. V.: J. Gen. Chem. USSR, *38*, 2551 (1968)
160. Green, B., Garson, L. R.: J. Chem. Soc. *C 1969*, 401
161. Blecher, H., Pfaender, P.: Liebigs Ann. Chem. *1973*, 1263
162. Geckeler, K., Bayer, E.: Makromol. Chem. *175*, 1995 (1974)
163. Geckeler, K., Bayer, E.: Liebigs Ann. Chem. *1975*, 1671
164. Morawetz, H.: in: Peptides — Chemistry, Structure and Biology, (Eds. Walter, R., Meienhofer, J.), Ann Arbor Sci. Publ., Ann Arbor-Michigan 1975, p. 385
165. Flory, P. J.: J. Am. Chem. Soc. *61*, 3334 (1939)
166. Bayer, E. et al.: J. Am. Chem. Soc. *96*, 7333 (1974)
167. Bayer, E. et al.: in: Peptides 1974, (Ed. Wolman, Y.), Wiley, New York 1975, p. 129
168. Mutter, M., Uhmann, R., Bayer, E.: Liebigs Ann. Chem. *1975*, 901
169. Mutter, M., Bayer, E.: in: The Peptides: Analysis, Synthesis and Biology, (Eds. Gross, E., Meienhofer, J.), Academic Press, New York 1979, p. 285
170. Becker, H., Lucas, H. W., Mutter, M.: Unpublished Results; Becker, H.: Diplomarbeit, Univ. Mainz 1980

171. Bayer, E. et al.: Tetrahedron *34*, 1829 (1978)
172. Mutter, M., Hagenmaier, H., Bayer, E.: Angew. Chem. *83*, 883 (1971); Angew. Chem., Int. Ed. Engl. *10*, 811 (1971)
173. Bayer, E., Mutter, M.: Nature (London) *237*, 512 (1972)
174. Bayer, E., Mutter, M.: Chem. Ber. *107*, 1344 (1974)
175. Mutter, M., Bayer, E.: Angew.: Chem. *86*, 101 (1974); Angew. Chem. Int. Ed. Engl. *13*, 88 (1974)
176. Arlic, P. J., Spegt, P., Skoulios, A.: Makromol. Chem. *43*, 106 (1961)
177. Calleja, F. J. B., Keller, A.: J. Polym. Sci. *A-2*, 2151 (1964)
178. Calleja, F. J. B., Keller, A.: J. Polym. Sci. *A-2*, 2171 (1964)
179. Tadokoro, H. et al.: Makromol. Chem. *73*, 109 (1964)
180. Pillai, V. N. R., Mutter, M.: Acc. Chem. Res. *14*, 122 (1981)
181. Rahman, S. A., Anzinger, H., Mutter, M.: Biopolymers *19*, 173 (1980)
182. Toniolo, C. et al.: in: Peptides 1976, (Ed. Loffet, A.), Univ. of Brussels, Brussels 1976, p. 597
183. Jung, G. et al.: Chem. Ztg. *101*, 196 (1977)
184. Anzinger, H., Mutter, M., Bayer, E.: Angew. Chem. *91*, 747 (1979); Angew. Chem. Int. Ed. Engl. *18*, 686 (1979)
185. Anzinger, H., Mutter, M.: Abstr. Am. Pept. Symp. 7th., p. 100 (1981); Polym. Bull. *6*, 595 (1982)
186. Mutter, M.: Tetrahedron Lett. *1978*, 2839
187. Bückmann, A. F., Morr, M., Johansson, G.: Makromol. Chem. *182*, 1379 (1981)
188. Geckeler, K.: Polym. Bull. *1*, 427 (1979)
189. Hemmasi, B., Woiwode, W., Bayer, E.: Hoppe-Seyler's Z. Physiol. Chem. *360*, 1775 (1979)
190. Tjoeng, F. S. et al.: Biochim. Biophys. Acta *490*, 489 (1977)
191. Tjoeng, F. S., Tong, E. K., Hodges, R. S.: J. Org. Chem. *43*, 4190 (1978)
192. Pillai, V. N. R., Mutter, M., Bayer, E.: Tetrahedron Lett. *1979*, 3409
193. Pillai, V. N. R. et al.: J. Org. Chem. *45*, 5364 (1980)
194. Rich, D. H., Gurwara, S. K.: J. Chem. Soc. Chem. Commun. *1973*, 610
195. Rich, D. H., Gurwara, S. K.: J. Am. Chem. Soc. *97*, 1575 (1975)
196. Rich, D. H., Gurwara, S. K.: Tetrahedron Lett. *1975*, 301
197. Pillai, V. N. R., Mutter, M., Bayer, E.: Paper under preparation
198. Colombo, R.: Hoppe-Seyler's Z. Physiol. Chem. *362*, 1393 (1981)
199. Colombo, R.: Tetrahedron Lett. *1981*, 4129
200. Colombo, R., Pinelli, A.: Hoppe-Seyler's Z. Physiol. Chem. *362*, 1385 (1981)
201. Hagenmaier, H., Frank, H.: Hoppe-Seylers's Z. Physiol. Chem. *353*, 1973 (1972)
202. Maser, F., Mutter, M.: Unpublished Results
203. Ragnarsson, U., Karlsson, S., Sandberg, B.: Acta Chem. Scand. *25*, 1487 (1971)
204. Ragnarsson, U., Karlsson, S., Sandberg, B.: J. Org. Chem. *39*, 3837 (1974)
205. Bayer, E., Mutter, M., Holzer, G.: in: Peptides — Chemistry, Structure and Biology, (Ed. Walter, R., Meienhofer, J.), Ann Arbor Sci. Publ., Ann Arbor-Michigan 1975, p. 425
206. Mutter, M.: Int. J. Pept. Protein Res. *13*, 274 (1979)
207. Hagenmaier, H., Mutter, M.: Tetrahedron Lett. *1974*, 767
208. Mutter, M., Hagenmaier, H.: Angew. Chem. *86*, 163 (1974); Angew. Chem. Int. Ed. Engl. *13*, 149 (1974)
209. Leibfritz, D. et al.: Tetrahedron *34*, 2045 (1978)
210. Ribeiro, A. A. et al.: Biopolymers (in press)
211. Redfield, A. G., Kunz, S. D., Ralph, E. K.: J. Magn. Res. *19*, 114 (1975)
212. Frank, H., Hagenmaier, H.: Experientia *31*, 131 (1975)
213. Frank, H., Meyer, H., Hagenmaier, H.: in: Peptides — Chemistry, Structure and Biology, (Eds. Walter, R., Meienhofer, J.), Ann Arbor Sci. Publ., Ann Arbor-Michigan 1975, p. 439
214. Frank, H. et al.: in: Peptides — Proc. Am. Pept. Symp. 5th., (Eds. Goodman, M., Meienhofer, J.), Wiley, New York 1977, p. 514
215. Frank, H., Meyer, H., Hagenmaier, H.: Chem. Ztg. *101*, 188 (1977)
216. Jung, G. et al.: in: Peptides — Chemistry, Structure and Biology, (Eds. Walter, R., Meienhofer, J.), Ann Arbor Sci. Publ., Ann Arbor-Michigan 1975, p. 433
217. Heusel, G. et al.: Angew. Chem. *89*, 681 (1977); Angew. Chem. Int. Ed. Engl. *16*, 642 (1977)

218. Mutter, M. et al.: Paper in preparation; Becker, H. et al.: Makromol. Chem. Rapid Commun. *3*, 217 (1982)
219. Pillai, V. N. R., Mutter, M.: Naturwissenschaften *68*, 558 (1981)
220. Naider, F., Goodman, M.: in: Bioorganic Chemistry — Macro and Multimolecular Systems, (Ed. van Tamelen, E. E.) Vol. 3, Academic Press, New York 1977, p. 179
221. Scheraga, H. A.: Acc. Chem. Res. *12*, 7 (1979)
222. Nemethy, G., Scheraga, H. A.: Q. Rev. Biophys. *10*, 239 (1977)
223. Gratzer, W. B., Doty, P.: J. Am. Chem. Soc. *85*, 1193 (1963)
224. Auer, H. E., Doty, P.: Biochem. *5*, 1708 (1966)
225. Lotan, N. et al.: Biopolymers *4*, 239 (1966)
226. Ingwall, R. T. et al.: Biopolymers *6*, 331 (1968)
227. Goodman, M., Naider, F., Rupp, R.: Bioorg. Chem. *1*, 310 (1971)
228. Mutter, M., Mutter, H., Bayer, E.: in: Peptides — Proc. Am. Pept. Symp. 5th., (Eds. Goodman, M., Meienhöfer, J.), Wiley, New York 1977, p. 403
229. Goodman, M., Saltman, R. P.: Biopolymers *20*, 1929 (1981)
230. Mutter, M.: Macromolecules *10*, 1413 (1977)
231. Mutter, M. et al.: Biopolymers *15*, 917 (1976)
232. Toniolo, C., Bonora, G. M., Mutter, M.: J. Am. Chem. Soc. *101*, 450 (1979)
233. Bonora, G. M., Toniolo, C., Mutter, M.: Polymer *19*, 1382 (1978)
234. Toniolo, C. et al.: Macromolecules *12*, 620 (1979)
235. Mutter, H., Mutter, M., Bayer, E.: Z. Naturforsch. *B 34*, 874 (1979)
236. Mayr, W., Oekonomopulos, R., Jung, G.: Biopolymers *18*, 425 (1979)
237. Roth, W., Heppenheimer, K., Heidemann, E.: Makromol. Chem. *180*, 905 (1979)
238. Anzinger, H., Schmitt, J., Mutter, M.: in: Preprints of the IUPAC Symp. on Macromolecules, Vol. 2, p. 527, Florence 1980
239. Anzinger, H., Schmitt, J., Mutter, M.: Makromol. Chem., Rapid Commun. *2*, 637 (1981)
240. Bonora, G. M. et al.: Makromol. Chem. *180*, 1293 (1979)
241. Toniolo, C., Bonora, G. M., Mutter, M.: Int. J. Biolog. Macromolecules *1*, 188 (1979)
242. Toniolo, C. et al.: Macromolecules *13*, 722 (1980)
243. Toniolo, C. et al.: Gazz. Chim. Ital. *110*, 503 (1980)
244. Bode, K., Goodman, M., Mutter, M.: Biopolymers (in press)
245. von Dreele, P. H. et al.: Macromolecules *4*, 408 (1971)
246. Ribeiro, A. A., Goodman, M., Naider, F.: J. Am. Chem. Soc. *100*, 3903 (1978)
247. Mutter, M. et al.: in: Peptides 1980 — Proc. Eur. Pept. Symp. 16th., (Ed. Brunfeldt, K.), Scriptor, Copenhagen 1981, p. 660
248. Toniolo, C. et al.: Makromol. Chem. *182*, 2007 (1981)
249. Toniolo, C. et al.: Makromol. Chem. *182*, 1997 (1981)
250. Maser, F., Mutter, M., Toniolo, C.: Biopolymers (in press)
251. Chou, P. Y., Fasman, G. D.: Ann. Rev. Biochem. *47*, 251 (1978)
252. Tanaka, S., Scheraga, H. A.: Macromolecules *9*, 142 (1976)
253. Wünsch, E.: Angew. Chem. *83*, 773 (1971); Angew. Chem. Int. Ed. Engl. *10*, 786 (1971)
254. Bodanszky, M.: in: Peptides — Proc. Am. Pept. Symp. 5th., (Eds. Goodman, M., Meienhofer, J.), Wiley, New York 1977, p. 1
255. Meienhofer, J.: Biopolymers *20*, 1761 (1981)

Bei Fragen zur Produktsicherheit wenden Sie sich bitte an:
If you have any questions regarding product safety,
please contact:

Walter de Gruyter GmbH
Genthiner Straße 13
10785 Berlin
productsafety@degruyterbrill.com